STIKINE ODYSSEY

FROM ADVENTURE TO ACTIVISM WITH THE GREAT RIVER

PETER ROWLANDS

One Printers Way
Altona, MB R0G 0B0
Canada

www.friesenpress.com

Copyright © 2023 by Peter Edward Rowlands
First Edition — 2023

Featured Photography by Gary Fiegehen

Editing by Janet Gyenes

All rights reserved.

No part of this publication may be reproduced in any form, or by any means, electronic or mechanical, including photocopying, recording, or any information browsing, storage, or retrieval system, without permission in writing from FriesenPress.

ISBN
978-1-7752383-5-5 (Hardcover)
978-1-7752383-4-8 (Paperback)
978-1-7752383-6-2 (eBook)

1. NATURE, ENVIRONMENTAL CONSERVATION & PROTECTION

Distributed to the trade by The Ingram Book Company

Also by Peter Rowlands

Man on the Move: The Pete Friesen Story (2009)

Some Pigs Can Fly: Aviation Anecdotes (2018)

Grampa Pete's Pictures and Poems (2021)

for
my children
DAVID and STEPHANIE
and their children

Table of Contents

Preface ix
MAP 1: Stikine Watershed xii
  1: Falling and Flowing 1
  2: Plunging and Pouring 11
  3: Rising and Resounding 25
MAP 2: Upper Watershed 36
  4: Leaping and Lunging 37
  5: Chattering and Clattering 49
  6: Splitting and Splicing 59
  7: Singing and Soaring 71
MAP 3: Grand Canyon 82
  8: Stretching and Striding 83
  9: Dashing and Dancing 97
  10: Resting and Reflecting 107
  11: Surging and Swerving 119
  12: Bouncing and Barging 127
MAP 4: Lower Watershed 142
  13: Swishing and Swirling 143
  14: Raging and Roaring 155
  15: Pushing and Probing 161
  16: Settling and Surrendering 171
  17: Mixing and Merging 181

Author Notes 187
Appendix A: Voices Passed 191
Appendix B: Declaration of the Tahltan Tribe, 1910 193
Appendix C: Rivers of Canada 195
Gary Fiegehen: Photography Notes 197
Acknowledgements 203

Preface

Some years after I last attended a meeting at Friends of the Stikine Society (FOS), its managing directors, Ann Jacob and Stan Tomandl, mailed me a package containing six scrapbooks of newspaper clippings pertaining to resource management issues in the Stikine Watershed over a ten-year period beginning in the late 1970s. Containing relevant articles from periodicals in southeast Alaska and the interior of British Columbia as well as from the major newspapers in Vancouver and Victoria, it's a comprehensive collection that barely misses a beat on the timeline. Although its primary focus is on BC Hydro's proposed five-dam megaproject, the clippings do provide ample insight into the political arena while introducing mining and forestry initiatives on top of the annual fish wars.

Treasure troves of information, these six scrapbook albums will be offered to the museum and archives of British Columbia in the near future. Meantime, with skis and paddles hung on the wall in deference to new leg joints and old muscles, long-simmering thoughts about maybe, one day, possibly writing a story on the subject have percolated to the surface. Unfortunately, by the time my pen came to hand, no one, not even Ann and Stan could remember how, and from whom, the scrapbooks had arrived. All subsequent leads and best hunches have proven fruitless. Fortunately, my ancient filing cabinet stash of paperwork produced a healthy collection of FOS newsletters—many but not all, covering the last twenty years of the twentieth century.

Maggie Paquet, who was the FOS newsletter editor for much of that period, came onboard with her stash of paperwork and we began connecting dots. Taking time from her other projects, Maggie was able to analyze and organize my scribblings and she was instrumental in getting this vehicle roadworthy. From years of friendship and constructive conversation came the title for this book. Thank you, Maggie.

Another friend and colleague, Gary Fiegehen, stepped up immediately and offered unrestricted access to his boatloads of Stikine photographs—many appearing here are grey-scaled copies from his 1991 book, *STIKINE: The Great River*. These big-picture images provide an excellent overview of the watershed and give proper context for my story. Thank you, Gary.

My aim in writing this book is twofold. While wishing to celebrate the inner and outer beauty of this one river, I also hope to show how commitment to the public process is essential for management of our natural resources. The first part is easy. Gary's images capture the watershed's big picture while my little snapshots (old and tired though they be) follow adventures on the river. Surfacing sporadically into the text without warning are lines from the poem *STIKINE* (my light-hearted attempt to capture the entire river on paper without using the same pair of alliterating present participles more than once)—attempting to keep the river's voice alive through mountains of paperwork and acres of verbiage; for the same purpose, selections of these paired participles are used as somewhat relevant chapter headings. For reader orientation, maps, endnotes, and a photo index are included. The second part of my mission is less easy. Ultimately, I hope my personal journey—seen as a thread through woven fabric—helps identify the varied strands of a highly complex process. Despite fairly rigorous elimination, copious amounts of administrative reportage remain present in reflecting the tough slogging of any resource management issue—easily ignored at reader's discretion. Similarly, many of my personal rants and ravings are retained in order to reflect the temper of the times—forgiveness may be necessary. While honouring people of the story who have since passed (Appendix A), I hope river lovers, history buffs, and the environmentally conscious can all find some pleasure in the journey.

Welcome aboard!

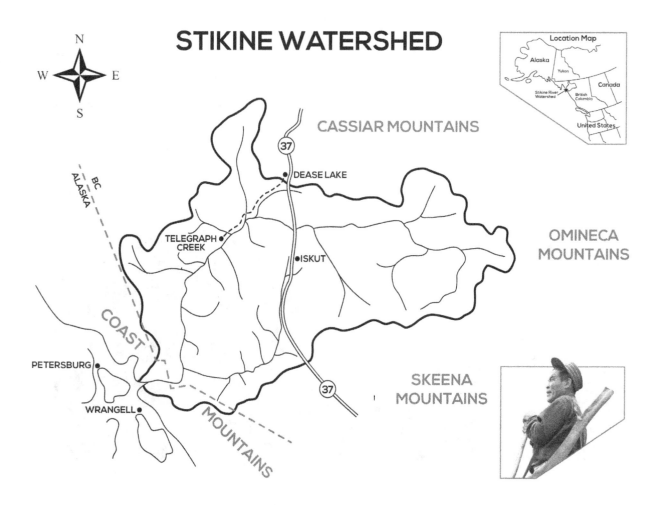

1
Falling and Flowing

Highway 37 was but a rough stretch of gravel when we motored north in September 1979, two youngish men heading for a paddling adventure in the wild northwest corner of British Columbia. With me was fellow Vancouver-based airline pilot Hal Marsden, who had become addicted to high mountains and white water while cutting his teeth on the South Nahanni, Back, and Coppermine rivers. By contrast, my slightly advanced age and lesser experience qualified me as a thirty-five-year-old wannabe with just enough knowledge to be dangerous. Riding along with us on the roof rack of my trusty Landcruiser (named Brutus) was our default paddling device: a heavy but durable seventeen-foot Grumman aluminum canoe known as *Dimples*. The allure of elementary school geography had combined with recent years of aerial observation to fire a passion impossible to ignore. Out in the middle nowhere between Terrace and Whitehorse ran 640 kilometres of wild river too beautiful to resist—*Stikine*.

The scene at the float-plane dock on Eddontenajon Lake, about 240 kilometres south of the Yukon border, brought a sudden sense of authentic Canadianism: backdropped by wild verticality and warmed by reflected sunlight off shining waters, a dozen or more humans idled among tons of boxes, bags, and boats being loaded and offloaded from a solitary yellow float plane that came and went as the day progressed. Hikers, paddlers, hunters, and prospectors—we were all in the same time zone and it had no clock. Unlike the feel of southern highways, there was no sign of haste or panic, only warm smiles and friendly banter along the length of the dock. Although we were all anxious to get where we were going, it was impossible to be in a hurry.

After lashing our old and wrinkled Grumman canoe to his port-side float strut, our calmly intent loadmaster arranged us and our gear inside his DHC-2 Beaver before pushing off the dock and battery-starting its 450-horsepower Wasp Junior radial engine. Our one-hour charter flight took us east-by-southeast over a southern portion of Spatsizi Plateau Wilderness Provincial Park to the upper reaches of the Stikine River where we were deposited on the shore of Tuaton Lake along with our canoe and most of our gear. Being a relatively inexperienced newcomer with multitasking inadequacies, I had inadvertently left my brand-new waterproof camera box on the departure dock. The cheechako had arrived.

The next day, my appreciation for our pilot and the unwritten "Law of the North" increased big time when, with slight diversion from another charter, he delivered my camera box to our beach while we were away hiking. No relayed messages and no charge; it was just done. Several subsequent experiences would confirm Murray Wood to be a well-proven pilot and a well-regarded person throughout the Stikine watershed community. Thank you, Murray.

Under partner Hal's leadership, our canoe trip began with a long walk. After securely stowing *Dimples* in the lakeside buck brush, we began a nine-day trek that took us west over a slight divide to Klahowya Lake in the Spatsizi River drainage before heading south and back east to explore the very upper reaches of the Stikine branch. Along the way, we wandered park-like alpine valleys where the only signs of humanity were the slightly worn horse trails of guide outfitters following in the footsteps of moose and caribou. A lesson soon learned: keep your hiking boots on when wading across shallow streams. Not only does that method offer superior footing while helping break in the boots to your foot shape, it prevents the need to use your Swiss army knife for surgical extraction of small pebbles from under reformed heel blisters. Lesson corollary: include strong pain killers and ample antiseptic ointment in the first-aid kit.

From our Tuaton camp at 1,341 metres above sea level, we scrambled up steep ravines and white-knuckled a ledge or two while ascending several headwater peaks above seven thousand feet, including Mt. Thule and Mt. Umbach, as well as our map's unnamed Peak 7429, which has been "Mount Marsden" in my mind ever since. While my blisters and anxiety were resting in camp, Hal scaled that pinnacle for a second time in the same day. This rare moment of inactivity gave me the space to indulge in retrospection, which heightened my appreciation not just of my surroundings but my place in it. Yes, Miss Johnson's wonderful Grade 7 teachings of 1954 did ignite my curiosity for this part of our country, a place she had never seen, one where names of its rivers and mountains are enticing in themselves—Bella Coola, Babine, Skeena, Kechika, Nechako.

My first look at the area had come from the cockpit of a WWII-vintage B-17 bomber in the summer of 1969, during a high-altitude photomapping assignment for Kenting Aviation. My appreciation subsequently blossomed with five years of flying the BC District as a B737 first officer with CP Air. Now camped in that high country and drinking meltwater straight from the mouth of a mountaintop glacier, I began to think about the role of this small tributary in a greater system. Unbeknownst in the moment, those first tastes of the Stikine introduced me to a relationship that would soon deepen.

Among numerous higher-elevation headwater lakes in a 52,000-square-kilometre watershed, the five here—Happy, Tuaton, Hotlesklwa, Ella, and Laslui—form a closely connected chain, giving rise to the greater river's mainstem. From the air and on the ground, it is a sight of extreme beauty—a twenty-kilometre strand of crystal-clear waters in a broad valley of yellow-brown grasslands, bordered by graceful slopes of variegated green, rising to formidable peaks adorned with a pearly string of early season snow. Within this snapshot, the Collingwood Brothers' guiding camp on Laslui Lake is barely noticed—the only human structure in the area and one of few in the entire upper watershed.

Attractive and expansive, this earthly space was suddenly special: no longer just a small corner of an airborne empire, it had become a broad tapestry rich in life. Though the immediate airspace was often busy with bugs, I could drink from a lake overflowing with tasty fish trying to leap into my canoe while songbirds and land mammals went about their business with little concern. Along with this big hit of true wildness came the enjoyment of unhurried contemplation in a state of semi-isolation, and with it, the realization of how nearly absent such moments had become in my "other" life. In the same way as night flying in a darkened cockpit allows for the study of stars, daylight rambles in alpine valleys introduce the study of tarns.

Aha! Of course! Depending on wind direction, some flatland ponds can provide source water for two different streams, ones that might eventually supply two different river systems, even ones that may ultimately flow to different drainages—the Pacific Ocean in one direction and the Arctic Ocean in the other. As such, the normally ocean-specific Arctic grayling had found its way into the upper Stikine drainage to mingle freely with the rainbow trout and Dolly Varden. Any Stikine tributary sharing a valley bottom with a tributary on a river flowing to the northern sea is a possible suspect, the most likely being the Kechika River on its northeastern divide.

For the first time in my life, I heard the sound of silence. As I recognized it as a tangible entity, I recalled childhood infections and reports of paper-thin ear drums that later ruptured during pilot training. Although listening and dancing have always brought pleasure, my musical talent had, since grade school, been well slotted into the "tin ear" category. Meanwhile, boundless beauty and uplifting spirit did not always include the sound of silence: one particular night's sleep was well-nigh impossible, denied by flocking ducks and geese spending the entire night in raucous discussion of their impending flights south. As necessary as migrations might be, it seems leaving this valley always makes for a difficult decision.

Another in a long series of blessedly warm September days gave us a pleasant paddle down the lake chain under a clear blue sky, soaking in the sun, with Hal catching fish and me taking pictures. Barely a breath of wind escorted us down Laslui as we watched the valley walls converge ever so slightly. Slowly but surely, a current developed beneath us and we were soon into a narrowing channel with quickening pace, being accelerated around a gentle bend before confronting a river full of rocks—dead ahead. We pulled out. Although leader Hal seemed little concerned, it was our first encounter with a real "rock garden" rapid and a wonderful opportunity to review the basics we had practised in North Vancouver's Seymour River weeks before. While paddles had always been comfortable in my hands, the dynamics of big-moving water and the techniques for dealing with it were but words in a book. A brief discussion soon identified the rocks of interest, and we planned our route through them. As we prepared to launch our white-water paddling vessel, White-Water Hal asked if Flat-Water Pete remembered the chosen route through the numbered rocks.

"Well … maybe not exactly," came my feeble response.

"Neither do I," laughed Hal. "Let's go!"

We were free and clear in no time. Retrospect and experience now combine to identify this challenge as an easy class II rapid at most. Although we'd soon encounter more difficult conditions, this particular one remains noteworthy by being my first. The long, hot summer of 1979 gave us extremely low water levels—great for training purposes. On subsequent journeys, this rock garden was barely a factor, sometimes totally invisible underwater. However, the seed for a growing river awareness had been planted. A short distance beyond, our fresh young river turned hard left and disappeared straight into a rocky ridge. We pulled out on river-right and made camp on soft ground in a grove of spruce. Fountain Rapids.

We had arrived at the first of several "points of interest" on the upper Stikine, as identified to Hal through overview reports by government agencies and from personal communication with a reported river user. Without a de facto trip report from any other paddler, this local knowledge of uncertain credibility became our guiding light.

Fountain Rapids is a magnificent stretch of running water in a deceptively charming setting. After the serenity of the sparkling lakes above, it's a sudden dive into boulder-strewn madness in the finest imaginable transition—a hard left turn down a steep, smooth incline for a sharp right turn into a narrow channel of unbridled enthusiasm. Wave upon wave of highly disturbed water is thrown into stacks of agitated froth, blending into a white carpet some five hundred metres in length. Extreme excitement. For the casual observer, it is a pleasure to behold and comes complete with all the appropriate sound effects. For a paddler or a swimmer, it is life-threatening violence. Slightly inland on river right, a short-kilometre portage trail ambles gently through boggy terrain and over a low ridge before dropping to a fine grassy launch site in a back eddy.

Our paddle on the upper Stikine in '79 was something special, and only by a handful of subsequent experiences would its significance sink in—two weeks of clear skies and light winds with good fishing in water warm enough to swim in. This upper-river trip was the one and only one when the heavens refused to rain for the entire duration. It was delightful. A light but lively current kept us moving forward, while riffles and small rock gardens kept our attention. Such introduction to river travel had me hooked from the beginning; with it came old memories of being late for elementary school because of endless fascination with wooden matchsticks racing curbside down meltwater gutters … bouncing along broad riffles and slicing through icy narrows … before disappearing over the big waterfall of death into storm drain heaven.

Some twenty-five years after last watching matchsticks diving into a drain came the thrill of diving into the Stikine River from the rocks below Chapea Rapids, about ten kilometres downstream of Fountain Rapids, and reason for the second portage of our 1979 journey. This one was motivated by two entrance ledges spanning the river and beyond my experience level; our distance from the nearest road and Hal's common sense were also major factors in the decision to carry around.

Notwithstanding the ability factor, the reality of having only one food supply in one canoe encouraged the wisdom of erring on the safe side, which we did. However, without any wisdom on the matter, we portaged on river-right for an awkward put in through a gap in the canyon wall. Perhaps we missed seeing or missed the significance of moose horns hanging from a tree on river-left, marking the trail for every subsequent carry well before the appearance of official signage. Meanwhile, we were on our own without a guidebook or written notes: all we had were third-person references to names on the map where the river might be troublesome. Between such places, or where we thought they might be, our topographical charts included numerous slash marks to indicate small rapids and faster water. Experience soon determined many of them to be misplaced or misrepresented, and we largely ignored them.

What a wonderful way to travel! Without notes or a GPS to distract us, we could take in the scenery and focus on the river's mood. Curious by nature and attentive by necessity, we experienced a tinge of the excited apprehension that must have accompanied explorers venturing into new territory. About an hour downstream of Chapea, we encountered Metsantan, a second small canyon (similarly known by its adjacent tributary creek) which was found to be an S-shaped kilometre of flat, fast water tickling twenty-metre walls (maybe some sandstone and pink-hued conglomerates) all jostled together in well-mangled blocks and columns exhibiting sharp corners and surprisingly smooth faces. Faces? Impossible to miss on river-right, standing aloof in a green meadow flecked with autumn gold, an isolated block of contorted rock gazed back with an assembly of weathered human faces etched in stone. Looking almost hand-carved, this naturally eroded rock wall displays a half dozen or more faces "sculpted" in front and side profiles, reminiscent of a totem pole. For me, it was a talisman from the past, a sign of good medicine on the river.

Freeing itself of these canyon walls, the young Stikine River completes a long, gentle left turn around Tomias Mountain on the southeastern corner of Spatsizi Plateau, from where it flows due north for its next thirty kilometres. In the meantime, here at the confluence with Metsantan Creek, we were back in civilization. Well, amid signs of civilization. The name Caribou Hide on the map shows where a lengthy and long-used trail fords a shallow section of our river in connecting Stikine country with Omineca territory to the east. Traversing the breadth of Spatsizi Plateau before following Toodoggone River eastward to the Finlay River and beyond, this ancient route would have qualified as a major highway for Indigenous peoples, carrying trade goods and human migrations while providing access for hunters and gatherers. Caribou hides would have been commonplace here as well as at the abandoned village up on Metsantan Lake, where remnant cabins still stand. These high-country habitations near the Stikine–Finlay divide, in the mixing zone of Tahltan and Sekani peoples, were not present when the first Europeans wandered through.

For Hal and me, cruising around the southeast corner in 1979, it was a far different experience than for European explorers. Ours was a two-week jolly in mapped territory eased by motorized transport at beginning and end. Almost too good to be true, we enjoyed low water and high sunshine while bouncing lightly through an endless array of small rapids and gentle rock gardens—singing and dancing in paradise. After a night at Caribou Hide, our paddling excursion saw another forty-kilometre day take us around the northeast corner of the plateau where the Stikine begins flowing in a generally westerly direction. Here at the bend, we stowed our paddles for a backward float providing lengthy looks to the east, up Chukachida River to Mt. McNamara (2,522 metres).

About thirty kilometres farther west, after a night at Sanabar Creek, we met Spatsizi River sidling nonchalantly in from the south with its collection of silt-laden tributaries from Spatsizi Plateau adding a few kilometres of parallel colour before mixing with Stikine's crystal-clear flow. The rusty hue of the incoming water is largely attributed to swaths of orange-red ferric oxide found in some areas of the plateau's 7,700 square kilometres known to colour the coat of many a mountain goat. The word Spatsizi translates from the Sekani language as *the land of the red goat,* whereby comes a plateau, a wilderness park, and a mountain, all bearing the same name as the river. Given the nearly equivalent length and volume of this upper-river sibling, its name could justifiably have been applied to the overall river system. Now doubled in volume with a greyish-green hue, the great river Stikine drives due west beneath the plateau's northern rim.

More rapids than chasm, Jewel Canyon is always a gem in some form or another. In the very low water of 1979, it seemed otherworldly with huge, rounded boulders appearing to rest on a smooth, flat surface sliding silently forward with barely a ripple and nary a trace of white. Excitement peaked with a comic photo of Hal executing a big draw stroke while pressed hard against a rock almost equal to his height. This picture would change.

About thirty kilometres later at the Pitman River confluence, we found a gorgeous campsite—a high, dry, and level spruce grove on a narrow spit of land with easy access. Downstream from here, the Stikine takes on a decidedly different mien: towering banks of clay and compacted sand tell the story of a prehistoric lakebed cut up and carried away by a long-enduring watercourse. It seems this remnant eighty kilometres of "gravel road" has a long story beginning with the volcanoes and lava fields to be found farther yet downstream.

Beggerlay Canyon is probably the finest site on the upper Stikine mainstem for the study of fluid dynamics. In the extremely low water of 1979, we were literally in it before we knew it. Drifting backwards in big sunshine and flat water with my camera clicking, a sudden small bump caused me to look back at Hal who was wearing a huge grin beneath his eye-balled movie camera. That was it. As never again experienced, the ledge drop at the entrance was miniscule, and the entire run of the canyon was as smooth and unrumpled as fast-moving water can be. Maybe ignorance *is* bliss.

Fast approaching the halfway point of its life and sporting a hefty body of water, the Stikine seems to hunker down and charge ahead with an increased level of vigour that is perceptible at any water level—a herd of wild rivers stampeding straight west. The Grand Canyon of the Stikine was the next big name on our map, and we weren't going down there in a canoe if we could help it. Even we newbies knew enough to get off the river at the Highway 37 bridge and start hitchhiking south to 'Tenajon for our vehicle. To access the lower river, it would be a 160-kilometre truck portage north through Dease Lake and then southwest to Telegraph Creek.

We hadn't yet seen it, but we knew it was there: the Grand Canyon of the Stikine—the eighty-kilometre midsection between 280 kilometres of upper river and 280 kilometres of lower river—the link between a vast network of interior headwaters and the more consolidated channel through the Coast Range. As we soon learned, the canyon is a narrow cut through a broad lava plateau created by once-active volcanoes on the Ring of Fire. We "old-time" student pilots easily understood how the canyon functions as a giant venturi tube, collecting water from a wide area to be funnelled and accelerated through a constrictive space—and accelerate it does! On the universal paddling code that rates difficulty on an increasing scale from class I to class VI, the Stikine's canyon is generally regarded as class VI+ and had yet to attract many paddlers, at least none that we knew of.

However, with a 40:1 stream gradient (40 vertical feet for every horizontal mile), it had long attracted the attention of hydroelectric dam builders such as BC Hydro and Power Authority which was now on-scene and on the job; the scant few motels and lodges were all busy with people and their vehicles. Having just experienced more than 250 kilometres of wild natural beauty in relative calm, it was a bit of a shock to suddenly encounter so many fellow humans in the noisy clatter of one small restaurant where we learned that hundreds of workers and their continuous helicopter shuttles were busily employed in the construction of townsite work camps adjacent to primary dam sites in the canyon. As a career airline pilot who enjoys helicopter skiing, I had no aversion to industrial development. However, for a naïve idealist enjoying his first taste of true wildness, some discomfort arose when I imagined two massive hydroelectric dams in Canada's Grand Canyon of the Stikine.

After retrieving Brutus, my old Landcruiser, we motored to Dease Lake for gas and groceries before heading toward Telegraph Creek on one of the world's most enjoyable drives. At least for some of us it is. For others, it is something to be feared. Not to be confused with the six lanes of traffic in a song of the same name, this Telegraph Road is about 120 kilometres of sometimes gravel, sometimes mud that follows the Tanzilla River down off the Arctic–Pacific divide, then arcs gently southwest across a broad plateau before getting down and dirty when nearing the Stikine River. With minimal signage and no side barriers, the steep S-turns wind down to a one-lane bridge that crosses the Tuya River coming in from the right. It was a beautiful sight made suddenly exciting by an outbound supply truck churning uphill in a cloud of dust on a lane and a half of gravel squeezed between a high rock wall on one side and a long way down on the other. On that first trip in, fading light and worn-out truck tires (one flat and one leaking) forced us to camp overnight beside the Tuya.

Back on the road, a steep climb onto a brief section of straight-and-level ridge top led to another series of steeply winding downhill traverses, providing our first glimpses of the Grand Canyon of the Stikine where it defines the southern boundary of Day's Ranch, itself part of a broad panorama.

Soon after, our narrow gravel road sidled precipitously up to the canyon's rim, where it became even more scenically interesting. With minimal signage and no side barriers, the road ventured straight onto a tapering peninsula of disorganized volcanic rock that dropped hundreds of vertical metres to raging rivers on both sides. Awesome!

At the pointy end of this already narrow ridge, a short walk across ugly solidified magma offered us a sweeping overview of a powerful confluence—the wild Tahltan River rushing in to meet the great Stikine River, backdropped by a vertical wall of rusty brown basalt crystallized into the image of a large wingspread bird presiding over the scene like an eagle or raven on a giant drive-in movie screen. With a cone-shaped remnant on its top, this familiar old landmark is known to many locals as "The Hat." Everywhere, the walls of both river canyons attest to eons of volcanic activity with numerous thick layers of basalt laid one atop another. Near the end of the rocky peninsula, Telegraph Road does a tight right turn and steeply descends a skinny traverse to cross the Tahltan River on a single-lane wooden bridge before angling equally steeply uphill on the other side while clinging to the wall of the main canyon. Tahltan Flats is the ever-known name given to the meadow below, now sporting a small collection of homes and smokehouses adjacent to its long-renowned fishing spot. A place with history, as we would later learn.

Meantime, riding on a leaky spare and carrying one flat tire, we somewhat naïve tourists motored into the community of Telegraph Creek where we became *somewhat* disappointed to discover no sign of a gas station or any other facility that might lend itself to the care of automobiles. An invitingly tight roadway took us steeply down the creek bottom to the riverfront shoreline where we found an old townsite of wood-frame buildings in various states of repair. Amid numerous long-abandoned abodes and a collapsed warehouse stood a well-preserved Anglican church and several comfortable-looking dwellings fronted by playful children and recycled furniture. Loaded with black and white film, my camera provided temporary distraction from any truck tire issues.

Alongside a string of garage-like outbuildings, the largest structure on the waterfront offered promise with its hand-painted sign above the door: RiverSong Café & General Store. The door was locked, and no one was home. At the other end of this short riverbank street stood a low-profile "office complex" with two doors, one of which sported an RCMP crest and a telephone number for use in an emergency; both of these doors were also locked. An adjacent building, however, had its garage doors wide open to a fine array of shop tools and engine parts, wherein a gentleman named Francis paused long enough in his busyness to counsel two naïfs on the realities of life in an isolated community and on the art of self-sufficiency. No, there are no automobile service stations west of Dease Lake, he schooled us; however, the RiverSong down the street usually has gasoline that works in car engines, and they would probably sell you some if you really needed it—and if they were open. Meantime, for tires, go back to the upper village and ask around for Henry; he might be able to help you.

Sure enough, Henry's house was well stocked with a wide selection of spare tires, one of which was the correct size for my Landcruiser and another one deemed "close enough." So far, so good. Air? No sweat! Through a line attached to the spark plug hole in one cylinder of his operating truck engine, Henry provided more than enough compressed air to get us back on the road. Bless you, Henry Vance! Along with special thanks to Francis for directions and to an unknown mother for the coffee and cookies, Telegraph Creek would always occupy a soft spot in my heart. On all subsequent visits, we would ride on four good tires and carry two good spares. Another lesson learned.

Our introduction to the lower Stikine River in '79 was the opposite of the near-idyllic experience on the upper section. Five days of in-your-face rain is the short story. Paddling the chop of a ten-knot current in opposition to a twenty-knot wind while hoping for a lunchtime lull in the action and a high and dry campsite at night was our daily ritual. Whenever so fortunate, we found welcome relief in one of the vacant snag-crew cabins where well-behaved field mice happily shared the warmth of a wood stove. The rumours we had recently heard of splendid scenery remained exactly that. We saw none of the fabled silver peaks of the Coast Range and very little of its renowned glaciers—other than occasional chunks of ice floating in the current.

Through rain and mist at the mouth of the Iskut, the sight of slicker-sodden fishers hauling humungous salmon from icy-cold brown water warmed our hearts and provided the first photo-op in several days. It also introduced us to the commercial fishery that had been initiated on the lower Stikine just that year and, although quitting early was a foreign concept for both of us paddlers, we accepted the fishers' kind offer of taking us and our canoe to Wrangell, Alaska, in their motorboat.

In retrospect, we foul-weather canoeists had probably provided their ideal excuse for a long-cherished trip to town. Oops! Far out on the mud-flat bay and within sight of our destination, it seemed the skipper of our wooden skiff became more interested in the attractions of civilization than in the function of his outboard motor. In a moment of unconscious rage, the starter pull-rope of his oft-failed engine disappeared into the briny deep … and a sombre silence fell upon his entire ship's company. It is often said we learn from our mistakes, and this is one such instance. The rain had stopped, and the wind had dropped: we could have paddled it on our own, although an extra day or so may have been necessary. As it was, a fortuitous boat tow got us humbly into the harbour at Wrangell, where US Customs and Immigration cleared us ashore. An overnight at the adjacent Stikine Inn saw Alaska State Ferries take Hal and our canoe downcoast to Prince Rupert while an available seat on a local air charter got me back upstream to Telegraph Creek for Brutus. Hal continued to Vancouver by air, and our well-complexioned aluminum steed was waiting for me in the ferry terminal compound at Rupert. The completion shuttle of any Stikine River paddle proved to be an exercise in itself; and although this journey had been successfully completed, the river was not left far behind.[1]

2
Plunging and Pouring

Understood to mean "The Great River," the name Stikine emanates from the language of coastal Tlingit peoples. As well as being an eternal food supply and a handy trade route to the interior, it is, by far, the largest and strongest river in their coastal neighbourhood where it flows through terrain of considerable magnificence. Draped with shining glaciers birthing headstrong streams, the Coast Mountains of the Tlingit are part of the single largest contiguous outcropping of granite in the world—a two-hundred-kilometre-wide band stretching 1,600 kilometres along the entire Pacific coastline of British Columbia.

To the northwest of the river, the extensive Stikine Icecap blankets the mountain crest while punctured by sharp peaks reaching three-thousand metres above sea level (ASL). Fed by innumerable streams while traversing several bioregions, the river's ultimate source is 640 kilometres inland on the high plateau known as Spatsizi—traditional territory of Tahltan and Sekani peoples—featuring broad valleys and more subdued heights of land. Between these two major upper and lower regions of the watershed is a massive volcanic complex on a north-south axis, approximately one hundred kilometres in length and thirty kilometres wide. With a broad caldera at 2,787 metres ASL, the map's Mount Edziza (known locally as Ice Mountain) is the high point of a broad geological formation containing shield volcanoes, calderas, lava domes, cinder cones and stratovolcanoes—the geophysical heart of the watershed. To the north, across Stikine's mid-river canyon lies a high-elevation range of mountains known as Level Mountain—another large volcanic complex exhibiting remnants of events on the Pacific Ring of Fire.

Initial research on the subject found the Tlingit's big river to be the bottom line of a 52,000-square-kilometre watershed of similar shape and slightly larger than the country of Switzerland, while being slightly smaller than its Skeena River neighbour to the south. In comparison, the province's largest river system, the Fraser, is 1,400 kilometres in length and drains 220,000 square kilometres, being itself only slightly smaller in area than the entire United Kingdom. While the Fraser River system occupies much of the province's southern half and contains most of its population, the Stikine drainage occupies a small corner of northern BC with only about eight hundred human inhabitants, most of whom are Indigenous people of Tahltan heritage.

Recent History

Along with European contact in the early nineteenth century came a flurry of fur trade activity and several gold rushes, large and small, all supported by a growing fleet of wood-burning paddlewheelers steaming upriver from the Pacific Coast. By the early twentieth century, the Stikine region had become a mecca for big-game hunters coming from far and wide in search of trophy animals within populations of grizzly and black bears as well as large ungulates. With still intact prey–predator relationships and vast tracts of undisturbed habitat, the area is often referred to as the "Serengeti of the North." While guide outfitters have long been established throughout the watershed, this specialty market may have peaked with the sizeable operation on the Spatsizi Plateau run by Tommy and Marion Walker during the 1950s. Tommy's book on the subject, titled *Spatsizi*, offers a unique introduction to the land and its people. This subregion's environmental significance was recognized by creation of Spatsizi Plateau Wilderness Provincial Park, and within it, Gladys Lake Ecological Reserve, in 1975.

Serious industrial development first touched Stikine in the 1950s when an asbestos mine opened in the Cassiar Mountains, 120 kilometres to the north. Later, in the 1960s, the mine's original access road—extending south from the Alaska Highway (the ALCAN of WWII vintage) near the BC–Yukon border—was gradually pushed farther south to Dease Lake on the Arctic–Pacific Divide where it connected with the road from Glenora and Telegraph Creek, which had been upgraded during WWII. Subsequently, that primary mine-access route was pushed a further 330 kilometres south from Dease Lake, aided by a cable-ferry crossing and subsequent bridge over the Stikine River, to reach tidewater at Stewart, BC, where oceangoing vessels could accept truckloads of asbestos ore directly from the mine.

Long known as the Stewart–Cassiar Highway, this gravelly old resource road officially became BC Highway 37 in the early 1970s when it was extended farther south again to connect with the Yellowhead Highway (Highway 16) east of Terrace. Adjoined by its 37A spur to Stewart, and eventually paved or seal-coated, Highway 37 has become a 725-kilometre tourist-friendly connector between Kitwanga, BC and Upper Liard in Yukon Territory.

As road construction activities prompted by the Cassiar mine were making a final push south in the early 1970s, another industrial incursion reached the Stikine from the southeast. Seen primarily as a route to northern resources, a planned connection with BC Rail's mainline at Prince George would extend as far north as Dease Lake. Along with the functioning asbestos mine in the Cassiar region and untapped coal deposits near the Spatsizi headwaters, the mountains of the central Stikine were known to contain significant amounts of copper and precious metals. No doubt, touristy passenger train service to Alaska was also part of the master plan when construction began on a track-worthy rail bed extending hundreds of kilometres through difficult-to-access and challenging terrain.

Surprise! Soon after this skookum new rail grade reached Dease Lake, the overall project was deemed unaffordable and was abandoned. Never used for its intended purpose, parts of the rail grade continue to serve back-country travellers seeking access to areas south and west of Spatsizi Plateau. Situated slightly east and upstream of the Highway 37 bridge, a multi-million-dollar untracked rail bridge over the Stikine remains a feature of interest for paddlers.

Several months after first passing beneath that rail bridge, another feature of interest caught my attention. March 24, 1980, found me at Robson Square Theatre in Vancouver for a presentation by Professor Irving Fox, from the University of British Columbia, who spoke adamantly against any damming of the Stikine River. It was a well-attended affair that brought joy to my heart and peace to my mind. For reasons yet to be fully understood, my heart and mind were finding common ground in distaste for the idea of damming the Stikine, and it was soon obvious here that many people not only shared concern for the river but were also unwavering in their intent to do something about it. Momentum had been building for some time, and a prior workshop had already brought together a broad spectrum of organizations and individuals to study the issue, discuss the options, and come up with recommendations.

Along with shared resolutions from that workshop, it was made abundantly clear that the concerns of local people must be kept foremost, with the people living "up there" having primary say in any management decisions for the Stikine watershed. Primary voices in any such discussions were obviously those of the Tahltan, the earliest known inhabitants of the watershed, who now represented about seventy-five per cent of its population, currently numbering about five hundred souls, many of whom were wholly or in part dependent on healthy fish populations and/or access to wild foods, including moose, caribou, and wild sheep. At this meeting and beyond, the unresolved Tahltan Nation land claim enshrouded all dialogue about Stikine watershed management like a morning mist—not always heavy, but always there.

Nevertheless, this public information meeting morphed seamlessly into a pep rally. Loose ends such as me were gathered up and sent forward. The Stikine River was overwhelmingly considered too valuable to dam. Hot damn, hallelujah! In support of Residents for a Free-Flowing Stikine, an action group based in Telegraph Creek, a Vancouver-based action group called Friends of the Stikine (FOS) had recently been formed and was now accepting applications for membership.

My initial Friends of the Stikine meeting took me to the home of Rosemary and Irving Fox on Vancouver's west side, near the University of British Columbia, where Irving was a professor at the School of Community and Regional Planning and at Westwater Research Centre for which he had been founding director. With a similar academic bent, Rosemary was already a well-recognized stalwart in the Sierra Club of Western Canada. Whenever forced to speak, Irving seemed to know what he was talking about. And it was impossible not to love Rosemary. Her Julie Andrews persona carried a spirited will too strong to mess with—a chainsaw wielded by a velvet glove. After having monitored BC Hydro's development plans for the Stikine River over several years, this couple, in conjunction with the Sierra Club of Western Canada, had initiated the referenced workshop that had prepared the ground for the March public presentation.

Also present at my introductory meeting of FOS was Tom Buri, another founding spark plug who had once homesteaded on the lower Stikine and was then a lawyer living in the Smithers area (where we later became neighbours for a time). Very recently, in support of this writing project, Tom provided an information package containing a comprehensive report on the proceedings of that original workshop. Though probably of more interest to history buffs than paddlers, I am including a synopsis of that workshop to highlight the complexity of the necessary root system beneath a growing campaign.

Beginnings

Review of the paperwork from Tom Buri shows the January 25–27, 1980, UBC workshop well organized and well attended by some four dozen participants including people from First Nations, environment groups, fishers, guide outfitters, recreation interests, lawyers, and academics, along with observers from BC Hydro and relevant federal and provincial government agencies.

As an introduction paper by Professor Fox clearly stated, the workshop had a defined purpose: "Through hard work and conscientious effort on the part of all participants, it should be possible to arrive at a specific program for marshalling widespread public support for a course of action that will preserve the Stikine as a free-flowing river." No, this was neither a fact-gathering exercise nor a debate upon the merits of any proposed development. The province had no established medium for assessing ecological impacts across a wide-angle view and had seemingly little interest in doing so. Industry had already decided there was no reason for *not* damming the Stikine River. The fight was on.

Armed with sixteen background papers, workshop participants separated into three discussion groups to address the issue. For the interested spectator, these papers are themselves treasure troves of information covering factoids of natural history as well as proven social impacts and potential legal implications. The Tahltans and their unresolved land claims (Appendix B) were prevalent in many a discussion. Whether directly referenced or not, the First Nations land issue wove a difficult-to-define sub-theme through the entire symphony. In one paper, Tom Buri offered this synopsis:

> The Tahltan people have used and occupied the Stikine watershed from time immemorial. They have laid claim to this territory from the time of first contact with the whites back in 1834. The tribal declaration issued in 1910 clearly sets forth the basis of their claim, and although they have been subsequently relegated to a few small reserves unilaterally set aside by the provincial government, the Tahltans have never accepted these reserves as settlement of their land claims, nor have they abrogated the claims in any other way …. A development proposal such as the Stikine hydro project may suddenly bring this claim to the forefront and provide an incentive for negotiations as

with James Bay and Mackenzie Valley. The threat of legal action and delays seems to be the only reason governments must settle land claims.

From this point of view, we can understand why native groups have been less than enthusiastic about conservationist efforts to stop these developments. It is not that the native people are opposed to protecting wilderness lands; they are afraid that a successful conservationist lobby would remove the only incentive government has to settle their claims. Over the years, the Tahltans have watched the continued erosion of their territory by unilateral government action. The two parks already established in the area did not address the issue of native rights. The recent alienation of agricultural lands in the Telegraph Creek area has placed some of the most valuable land in the hand of private owners. The Dease Lake extension of the B.C. Railway and the Stewart–Cassiar road are typical of many intrusions which have tremendous secondary impact on the native way of life …. Certainly, a great deal of trust and confidence will be required for the mutual support of native concerns and conservation interests.

Several other papers stressed the importance of fish as the crucial link between people and their river. In this case, Stikine was known to support all five species of Pacific salmon as well as rainbow trout (steelhead), cutthroat trout, and Dolly Varden char (all salmonids), and all supporting commercial and recreational fisheries. The Stikine's spring run of spawning eulachon was one of the largest on the northwest coast, and its estuary in Alaska was known to harbour fifteen-plus species of fish, from which several hundred thousand pounds of flounder and sole were trawled annually. Also, three species of crabs provided about 50,000 pounds while three species of shrimp provided about 100,000 pounds per year to the local fishery. Prawns and three species of clams were also present.

While the proposed hydroelectric dam sites were seen to be somewhat above the reach of spawning salmon, the extent of the dams' effect on migrating fish and their habitat was limited to pure speculation. Though water levels would probably vary out of sync with the natural cycle, as would its temperature, siltation, and chemistry, the degree of such changes and their effects on the fish would be but items of conjecture without large-scale research and site-specific studies … only possible if the dams were in place (à la catch-22). The effects were similarly unknown for the upper-river fish expected to be caught between the reservoir impoundment and a pounding through penstocks and turbines. Meanwhile, there seemed little impetus for spending taxpayers' money on expensive studies simply to validate what BC Hydro was already saying.

One of the more colourful submission papers came from wildlife biologist Dr. David Hatler regarding estimated effects on wildlife. While outlining anticipated alterations in habitat and migration routes for primary ungulates (moose, wild sheep, mountain goats, and caribou), as well as for wolves and two species of bears, both above and below the dam sites, Doctor Dave was quick to point out how little was known about the animal populations in question. Prior to 1974, inventory and management for all species in the northern half of the province was the responsibility of a single biologist in Prince George. Although subsequent posting of a second biologist in Fort St. John had spread out the paperwork, all wildlife surveys remained more extensive than intensive. While public interest had prompted attention to the ungulates of the Spatsizi with regard to park management and possible road access, there had yet to be any detailed wildlife survey for the lower river; moreover, there had never been a study designed to evaluate the importance of lowlands to wildlife anywhere in the Stikine.

In an aside to conservationists, he strongly urged a more philosophical appreciation for big-game trophy hunting; otherwise, much effort toward "preserving nature" is unwittingly contributing to the devaluation of wildlife and wilderness by assigning fewer animals to a smaller space for so-called "non-consumptive" use. Trophy animals require plenty of space, something that is good for all.

In referencing concerns for approximately two hundred mountain goats inhabiting Stikine's Grand Canyon, Dr. Hatler suggested the BC Hydro biologists on site had the most accurate information. The fact that their budget for one species within a 50-kilometre range was far larger than his government department's annual budget for all species in the northern half of the province was another matter. In summary:

> Wildlife values of greatest concern in the Stikine hydro project area are moose wintering areas within the impoundment area and below the dams, integrity of the low-elevation mountain goat population in the Grand Canyon area, possible annual migrations of Spatsizi's caribou across the proposed impoundment area, possible nesting of anatum peregrine falcons within areas subject to disturbance during construction, and opening up of new country to human visitation and use via construction roads and transmission lines with no increases in management or enforcement capability. The above list is not in any particular order; the last mentioned may be the most significant, and the caribou problem may also be very important, especially since it does not appear that it is presently being considered.
>
> In terms of actual hard data, BC Hydro will hold pretty well the full deck. Until their cards are on the table, we will not know what we are dealing with and even then, we will only be able to question terms of reference, assumptions, and possibly interpretations. The data they want us to see will be all we have to work with. While I do not question either the competence or the motives of the Hydro biologists, I strongly question the public policy that places the proponent fox in unaided watch over the project chickens. Perhaps <u>that</u> is what we should be working on.

Ω

Mr. Lew Green, a consulting geologist, prefaced his overview of mineral potential in the Stikine watershed with this quote: "Predicting the mineral potential of any area is as risky as promising perfect weather for the day of the Sunday School picnic. In the former, everything can be turned topsy-turvy by a new surge of exploration. This can be set off by a change in metal prices, such as the recent increase in gold, or discovery anywhere in the world of an important new deposit that might be duplicated in the area under study."

Since the early days of discovering gold on river bars in 1861, the value of mineral production in the region had been considered negligible, despite a whirlwind of exploration made possible by the use of helicopters after 1950. Other than significant coal deposits in the Klappan headwaters region, the entire upper Stikine had *low* potential for mineral extraction, while values increased to *moderate* around the canyon section before hitting the *high* mark in mountains of the Coast Range. Considered to have been intensively surveyed and known to contain sizeable deposits of copper and molybdenum as well as some precious metals, the area was not seeing much activity at the time because of low copper prices and high development costs attributed to inflation. While the proposed hydroelectric development was seen to present negligible effect, the ever-present access issue remained front and centre. The transport of ore from any mine within the so-called "Golden Triangle" between the Iskut and Stikine rivers would require tunnels of roadway to reach Highway 37 to the east or barges of river traffic downstream to Wrangell for ocean transfer. Neither option appeared particularly inviting or possible at that time.

Nevertheless, two other access options were already leaving paper trails. The order-in-council that created Spatsizi Wilderness Provincial Park contained a provision to allow construction of a mining access road from the Toodoggone River area (east of the park) up the Stikine branch and through its upper lake country to exit one of its uppermost tributaries enroute to the BC Rail grade—a frightening prospect for anyone who had experienced that corner of the watershed. Presumably, for as long as the Toodoggone-area mine remained out of production and the railway failed to complete, there existed little danger of this road corridor being developed. Meanwhile, discussions had already been taking place between the governments of BC and Alaska regarding a possible transportation corridor from tidewater along the lower Stikine mainstem to proceed up the Iskut Valley and connect with Highway 37. The western section of such development would compromise the Stikine–LeConte Wilderness area in Alaska, a designation recently ratified by the US House of Representatives, but not yet by the Senate. Another scary thought for many.

As indicated in one paper by Professor Fox, transportation corridors were a subject of widespread interest. On the assumption of the dams on the Stikine, Iskut, and Liard rivers going forward, BC Hydro was actively investigating four transmission line alternatives for getting its power south: a) east of the Rocky Mountains; b) the Rocky Mountain Trench; c) the BC Rail route; d) the Stewart–Cassiar Highway (#37). Along with assimilating the immensity of the project itself, concerned minds were having to integrate the vast scope of its infrastructure. The generally accepted forecast for ever-increasing cost of petroleum-based energy supplies was making hydroelectric power seem an even more valuable resource. Increased demand for such could be further accentuated by the potential market in the United States. With a population greater than all of Canada, Pacific states in the US might potentially experience (somewhat) similar increases in the price of gas and oil. It takes little imagination for any developer to envision an electric power grid extending from Yukon to southern California being fed by river generators in Yukon and southeast Alaska as well as from all British Columbia.

Plotting of all potential dam sites and transmission lines on a map of known mineral reserves brought cause for alarm. If our business-as-usual ethic prevailed, a series of incremental and disjointed responses to demands for development could conceivably carve up the north as we know it, reducing wilderness and wildlife to mere remnants—no more Serengeti of the North. Unless our collective understanding changed, northern British Columbia could be nickel-and-dimed away. Significant policy revisions were being deemed necessary if we were not to lose all our major rivers to hydroelectric development and all of our wildlife to pollution, poachers, and trophy hunters. Furthermore, it was recognized that agreement for dams on any one river was unlikely to buy protection for any other river: the Liard wouldn't be spared by damming the Stikine.

The Stikine's beauty and free-flowing nature had been documented long before my fateful attraction. A 1973 report for Parks Canada by Dr. M. A. Roed, titled *Theme Study of Geologic and Related Features and Phenomena,* describes twenty-seven outstanding features in Canada's three westernmost provinces. British Columbia's Grand Canyon of the Stikine was the only major canyon to make his list. In submitting his report to Parks Canada, Dr. Roed had included a covering letter in which he opined that the Grand Canyon of the Stikine should have been included in Mount Edziza Provincial Park. Coincidentally, the 1973 Parks Canada survey team had recommended setting Stikine aside as a designated "wild river."

Parks Canada had also identified the Grand Canyon of the Stikine and Spatsizi Plateau (including the Stikine headwaters) as areas of national significance. BC Provincial Parks & Outdoor Recreation Division had listed Stikine as one of several provincial rivers worthy of consideration for inclusion in the Canadian Heritage Rivers System (CHRS) being planned by federal and provincial jurisdictions. While always difficult to assess for dollar value, recreation use on the Stikine had been increasing. It was estimated that organized and private trips, rafting and canoeing, had accounted for approximately one hundred paddlers during the previous summer of 1979. By looking at the records and by having seen no other September paddlers, it seems our party of two (Hal and me) had fallen outside of collection parameters.

On the subject of possible options for river protection, a paper by Tom Buri provided clarification of several major factors on the legal aspects of preservation. Obviously, he proffered, some form of counterproposal for use of the river would be the most effective opposition to BC Hydro's plans; and given the uncertainties around future park creations and the immediate concern for the Stikine's Grand Canyon, the nascent Canadian Heritage Rivers System (CHRS) might be the answer. If modelled on the US *Wild and Scenic Rivers Act,* Canadian heritage river status would certainly prevent hydroelectric development. However, caution must be exercised if importing the US model, which offers no protection during the study phase and which, after designation, only provides protection for water-based projects: all logging, residential, and industrial developments can only be controlled by formal acquisition of corridor lands.

Surprisingly, Canadian provinces enjoy far greater jurisdictional autonomy than their State-side counterparts. Given the current balance of power between Ottawa and the provinces, it was unrealistic to imagine a CHRS act that would give Ottawa the exclusive power to unilaterally designate which rivers were included in the program. Even with absolute federal control, a ban on federally funded water projects would hardly inconvenience BC Hydro; similarly, any prohibition on National Energy Board licences would not apply until it came to exporting power.

By crossing an international boundary, the Stikine River adds another factor to the management equation. The *British North America Act* assigns title of the land (and riverbeds) to the provinces, which also gain management control of public lands through given powers over "local works and undertakings." The nation retains control over international relations, including navigation and shipping, coastal and inland fisheries, as well as responsibility for "Indians" and lands reserved for Indigenous peoples. Clearly, the basic authority over a river such as Stikine lies with the province … unless some tireless retro-lawyer wishes to redefine the clause "for the general advantage of Canada." As such, the province has the power to initiate legislation dealing with its rivers, and the federal government has veto power over any aspects of that legislation that may infringe upon its areas of jurisdiction.

The *National Park Act* requires transfer of land title from Crown provincial to Crown federal, making creation of a national park less than inviting for the province. While a CHRS program might not require transfer of land title in seeking provincial guarantees for free-flowing water quality and shoreline development, it would necessitate negotiations for allocation of development and management costs, assuming the province was willing to forego all hydroelectric dams and other future developments. The *Provincial Parks Act* requires no land title transfer and little or no federal participation. However, we were well aware that new provincial parks such as Spatsizi were created by order-in-council and, as such, can be summarily dispatched to the scrap heap as quickly as they were created.

While watching the possibility for a one-of-a-kind federal-provincial joint-management plan on Stikine slide under the bridge came the fantasy of a BC provincial rivers program that may one day recognize the value of preserving free-flowing natural resources. In the meantime, the unfortunate irony of the situation was that the government with the primary legislative power to establish a river park was also the government with the least incentive to do so. We could only hope for federal initiatives that might induce cooperation in time to forestall hydroelectric developments on the Stikine River.

On January 27, 1980, at the University of British Columbia, several dozen concerned citizens concluded a three-day search for ecological sanity by reuniting in a final plenary session to summarize findings and discuss future options. Following lengthy discussions about ways and means for information sharing and attracting support, a press release declared the participants unanimously opposed to the construction of dams on the Stikine and agreed that it should be preserved as a free-flowing river while acknowledging the Grand Canyon of the Stikine as one of Canada's scenic and geologic wonders. In calling for such preservation, workshop participants stressed the importance of not jeopardizing Indigenous land claims. To address longer-range issues, a committee under the chairmanship of Professor Irving Fox was formed for the purpose of organizing a task force to develop policies for wise use of northern resources. Also announced was establishment of a citizens' committee to monitor the Stikine and to recommend options for its management—a northern section in Telegraph Creek under the chairmanship of Wenda Marion and a southern section in Vancouver under the chairmanship of Rosemary Fox. After the requisite number of WHEREAS clauses came RESOLUTIONS:

> **Resolved** by the participating citizens and citizen groups that the Stikine River be maintained as a free-flowing river and the BC Government instruct BC Hydro to immediately halt all hydroelectric studies and developments on or related to the Stikine watershed.

Resolved by the participating citizens and citizen groups to strongly urge the Government of British Columbia to immediately begin a Public Inquiry into BC's provincial energy policy.

Resolved by the participating citizens and citizen groups that the Provincial Government be requested to introduce forthwith a policy that any organization, whether public or private, such as BC Hydro, which proposes to use any natural resource, must pay to the Government or at its direction a percentage of its estimated capital cost to be used for the preparation of adequate and independent studies into the environmental and social impacts of the project, and that the assignment and use of these funds be directed by a committee consisting of the relevant Government agencies, the project proponent, as well as local and private-interest citizen groups.

Resolved by the participating citizens and citizen groups that BC Hydro be instructed by the Government to release forthwith to the public all its data and studies on proposed hydro development, including the Stikine River, whether draft, preliminary, or final … so that it can be established that all relevant matters are given proper study; and to continue to release to the public all such studies as soon as they become available to BC Hydro.

Inspiration

As previously disclosed, I did not have this workshop information at the time of my enlistment in Friends of the Stikine, not that it was under lock and key or otherwise unavailable, but it was paperwork and I had no desire to delve into the details, Top Secret or otherwise. A foot soldier doesn't need to know about plans and strategy—it might be confusing—just point him in the right direction and tell him when to open fire.

Soon after my introduction to FOS, a good-looking guy named Wade Davis knocked on my basement suite door; he wanted to talk about the Stikine. Okay by me. As I soon learned, Wade had recently arrived in town on a round of wide-ranging interests. For starters, as a former park ranger in Spatsizi Plateau Wilderness Provincial Park, he had come to know much about the area and about its people. In the sharing of informational gems, his sincerity was inspiring and his demeanour refreshing; my sense of commitment to the cause grew even more solid. Unfortunately, this mini workshop ended far too soon. Wade was on his way to South America—something to do with plants and people in the Amazon. For me at the time, ethnobotanists were an unknown species. Occasionally reconnecting over the years since, including visits at his beloved Wolf Creek family abode, we would keep somewhat up to date with each other's Stikine-related activities. Wade was and remains an inspiration.

HISTORY 101

For anyone who enjoys reading maps and paddling rivers, it is impossible not to appreciate the efforts of Samuel Black and his party of Hudson's Bay Company (HBC) explorers who traversed an upper section of Stikine country back in 1824. Though certainly not the first humans to experience the area, they seem the first to leave behind a written record of such an adventure by the journey log of Samuel Black. As was their custom, Black and his companions had dedicated a full season to the task, from break-up to freeze-up, pushing the limits of their endurance while pushing through unknown territory and inclement weather, accompanied, of course, by hordes of black flies and endless mosquitoes. Vital to this enterprise were Indigenous guides[2] to help with directions and translations as well as with stocking the larder. From their wintering place at Rocky Mountain Portage (near Fort St. John), they began by pushing, pulling, and paddling 120 kilometres up the formidable Peace River—the only river penetrating the Rocky Mountain divide—then, instead of turning south along the Parsnip branch that had facilitated Alexander Mackenzie's first-ever transcontinental crossing thirty years prior, Samuel Black was directed to turn right and explore the river's northern branch, which in the interim had been identified by fellow trader John Finlay.

After a strenuous and perilous ascent, Black and his party reached Thutade Lake at the headwaters of Finlay's River by midsummer of 1824. Having fought a ferocious current upstream for almost five hundred kilometres, they had unwittingly discovered the ultimate source of the 4,200-kilometre Mackenzie—Canada's longest river and North America's second largest river system. However, like his season, Black's work was only half done. From Thutade Lake, it was a 400-kilometre round-trip hike through wild and rugged terrain in search of fur trade possibilities. Along the way, he identified this river we know as the Stikine and later "discovered" another one directly north, which he named Turnagain. It was not a pleasant walk. The weather was uncomfortably cold and wet, and manpower desertions were common. Fortunately, aided by Indigenous people long accustomed to the land, Black's company of hardy adventurers made it "home" for Christmas. The fact that his boss, (Sir) George Simpson, had assigned him an exploratory mission in the highest terrain at the farthest edge of HBC territory invites a question about their relationship.

At the time of this journey, Black, a 44-year-old Scot and a former major rabble-rouser with the North West Company, was being assimilated with some difficulty into the culture of his new HBC employer. According to Simpson, "This Outlaw is so callous to every honourable or manly feeling that it is not unreasonable to suspect him of the blackest acts." Later, in his celebrated *The Character Book of George Simpson, 1832*, he described Black as, "The strangest man I ever knew. So wary & suspicious that it is scarcely possible to get a direct answer from him on any point, and when he does speak or write . . . so prolix that it is quite fatiguing to attempt following him."

Nevertheless, Black's ability and reliability were too valuable to ignore. After discovering fur trade possibilities to be highly unsatisfactory in the area of his northwestern exploration, he was given positions of increasing responsibility, leading to his appointment as chief factor in charge of the inland posts of the Columbia region while headquartered at Thompson Rivers Post (Kamloops). There, in 1841, Samuel Black was subsequently shot to death under curious circumstances … but that's another story.[3/4]

3
Rising and Resounding

The early days at FOS were easy to bear for the average foot soldier. Monthly meetings provided updates from our executive branch (Rosemary Fox) while allowing ample time for commiseration on the latest media coverage, or lack of it. A common thread found in newspaper reportage was a discrepancy in the amount and type of public involvement claimed by BC Hydro and by people of the Stikine. "We don't know what Hydro is going to do here," said Gordon Franke from the Telegraph Creek Band. "We get no straight answers from Hydro. They tell us one thing and they say something else in Vancouver."

The public utility was withholding its study results from the public. To date, $4.78 million had been spent on studies for a five-dam Stikine–Iskut project, with $31.7 million budgeted before 1985. Already, for fisheries concerns, it seemed likely that Hydro's stack of research would swamp any meagre offering from the federal Department of Fisheries and Oceans (DFO). In countering Hydro's claim of widespread (southern) public support, Tahltan Chief Ivan Quock of Telegraph Creek introduced Willie Williams, whose family run horse ranch is the only riverside habitation above the canyon. Along with its sightseeing trails, all twenty-seven acres would be inundated. As suggested by regional MLA Al Passarell (NDP, Bulkley–Stikine), it would be morally wrong to flood Canada's Grand Canyon.

The contrarian consensus of many northern workers came from Bruce Busby, editor of the *Interior News* in Smithers: his rambling diatribe against the rationale put forth by the Sierra Club's anti-dam presentation at Robson Square on March 24, 1980, maintained that greed was good if it was in the best interests of the greedy majority. "Professor Fox and his ilk seem not to want to address themselves to these realities, preferring instead the anarchy of their small conclave insulated by a kind of intellectual laziness and all made possible by the very industrial and commercial aggressiveness they oppose."

While the idea of anarchy continued to generate smiles within our small conclave, subsequent FOS meetings did introduce additional "intellectual laziness," such as Deborah and Tom Buri who had prepared background papers for the January workshop, along with the ever gracious and supportive May Murray of the Canadian Parks and Wilderness Society (CPAWS) who was already an established energy source within the local chapter and beyond. From time to time, idlers from other organizations would also lose their minds and drop by to shuffle papers and donate more of their unpaid free time (irony intended).

People of the Stikine were being motivated. SAVE THE STIKINE—STOP THE DAMS! was an appropriate call to action appearing on the front page of a newspaper-like newsletter published and distributed by Residents for a Free-Flowing Stikine (RFFS), the purpose of which, as outlined by Joe Murphy, was to introduce the BC Hydro project to a province-wide audience while identifying the implications for the people who live there.

A brief history: Following the discovery of large copper deposits in the area during the early 1960s, hydroelectric field studies were conducted by Brinco Ltd., a Montreal-based British company that had developed the Churchill Falls hydroelectric project in Labrador. Then, in 1972, Premier W. A. C. Bennett's administration placed a flooding reserve on 320 kilometres of the Stikine River to facilitate the provincial scheme. A change in government soon thereafter saw NDP Resource Minister Bob Williams veto Brinco's planned project on ideological grounds as well as on appreciation of the river's scenic qualities. When Brinco tried to regurgitate its plan under a subsequent Social Credit government, it was summarily shunted aside by a newly created Crown corporation—BC Hydro and Power Authority—which began reassessment of the Iskut–Stikine in 1976. Two years later, it pronounced a five-dam 2,690-megawatt power development to be economically feasible with a projected online date in the early 1990s. Interestingly, while the official position cited growing provincial needs as rationale, BC Hydro Chairman Robert Bonner was simultaneously advocating the export of power to the United States, with specific reference to the Iskut–Stikine.

The people who lived there were not impressed, especially when learning the generated high-voltage electricity would be unavailable to them. "We're not against northern development; we need it because of our high unemployment. But we don't like it when people come up from the south to use our resources and then export the net benefits back to the south," said Lorgan Bob of the Tahltan's economic development committee. Similar sentiments expressed by Rosemary and John Plummer, fishers, and homesteaders in the valley, referred to the impending dire effects of such a megaproject on their small community: alcoholism and suicide go hand in hand when people are alienated from their land. While strongly urging all cheap-power proponents in the south to also consider the needs of their referenced "few Indians, homesteaders, and fishermen," the Plummers emphasized the intrinsic value of unspoiled land and viable communities—they become more precious with time.

Hopes for a commercial fishery on the lower Stikine had taken flight the previous summer (1979) when BC Packers had anchored a brine-tank vessel just inside the Canadian border. Results of the season showed higher numbers of coho and sockeye salmon than anticipated despite the customarily huge allotment taken by fishers on the other side of the international boundary line. It was also determined that a higher than anticipated number of Stikine salmon were spawning in main channel sloughs and gravel bars. As local fisher Florian Maurer pointed out, regardless of any "high quality" dam management, these large numbers of "stemmers" would be adversely affected by the hydroelectric project, along with all hopes for a processing plant at Telegraph Creek, then being seen as a means for restoring a degree of self-reliance to the Tahltan community. Maurer concluded: "Interestingly enough, BC Hydro selected for its studies only such sites on the lower river where, in heavily silted glacial tributaries, no spawning can possibly occur. Apparently, federal Fisheries cannot allocate sufficient money to do studies of its own."

Recognizing the overall concern for maintaining natural balance in the Stikine watershed, Smithers-based botanist Rosamund Pojar summarized Dr. Hatler's previous report to the Sierra Club which had highlighted considerable lack of knowledge about very large wildlife populations. Although suggesting minimal impacts by dams and reservoirs on the regional flora, she gently cautioned about possible detrimental effects on the regional agriculture-friendly microclimate.[5] Expressing frustration about the general lack of government research in the public interest while BC Hydro is permitted to carefully and selectively release its information to the public, Pojar concluded, "The corporation's conclusions will thus suit their own ends ... and it is unlikely that an independent economic assessment can be made."

Being stonewalled by government agencies in his search for background information about decisions for pipeline construction, hydro rate increases, and names of petrochemical applicants, the *Vancouver Sun*'s energy writer, Don Whiteley, summarized his frustration: "The ministry's decisions are increasingly being made with information the public has never seen and will never see. Trends are being established to guarantee that an absolute minimum of information will find its way into the public domain."

It has often been said that too much democracy—inputs and opinions—can be a hindrance in the short term when action is required; however, the question arises here about the public's right to know the methods by which a defined public entity, funded with public money for the public good, conducts its business. This issue popped into view around BC Hydro's application for construction of canyon access routes from Highway 37: a southside road to explore for gravel deposits and a northside cat trail to transport drilling equipment deemed too heavy for helicopters. When barred from the decision-making meeting in Smithers, the northern citizens' group Residents for a Free-Flowing Stikine (RFFS) applied to the provincial ombudsman who advised that no requirement existed in provincial land legislation for such a meeting. Instead, it was suggested they call a public meeting to which Hydro and government agencies would be invited. As reported by Joe Murphy, RFFS spokesperson, it seemed the Ministry of Lands manager responsible in Smithers saw no value in such a meeting because local sentiment was already well known from letters of opposition he had received. Instead, he asked BC Hydro to make its applications public while he sought input from other resource agencies. As numerous public interest groups began inundating then Minister of Lands Jim Chabot with anti-access letters, RFFS anti-dam spokesperson Joe Murphy became an elected director of the Kitimat–Stikine Regional District.

In March 1981, without BC Hydro's participation in any public meetings on the issue, the Ministry of Lands, through its regional office, rejected Hydro's application for road access to the canyon on grounds of environmental risk and the availability of other options.

The response by BC Hydro's Chairman Robert Bonner was forthright and addressed directly to Minister Chabot:

> … inappropriate for your ministry to adopt the position that Hydro must hold public hearings as a prerequisite to approving our application…would cost Hydro and thus their [sic] customers … can seriously affect Hydro's ability to bring the development on stream at the appropriate time, **if approved by the proper regulatory agency** [emphasis mine]. The decision has been made on inappropriate grounds, on erroneous technical information, and without an appreciation of its importance and consequences. I therefore request within your ministry an immediate review of this matter at a high level.

While strongly urging the minister to override the wisdom of his ministry, it seems Mr. Bonner maintained presence of mind by using the "if" word in tactful regard to a process seen by him and others as fait accompli. Whatever the proper agency might have been, his choice of words may have caught the attention of the regional Lands' department as well as the minister himself. In any case, after launching its appeal, Hydro was directed to participate in public meetings before any decision would be rendered.

According to Moira Farrow's report in the *Vancouver Sun* (February 24, 1982), the first such event was a disaster. Held in Terrace and hosted by the local Jaycees, BC Hydro's presentation was allowed to take twice its forty-minute time allotment before public interaction was then cut short by strict limits on the number of questions allowed. Among the most vocal dissenters, lawyer Tom Buri, representing RFFS, decried the five ministry representatives present for sitting as representatives rather than presiding at the meeting: "Is this a hearing or a slick publicity performance by BC Hydro?" Also, FOS Director Bill Horswill strongly suggested Hydro's exploration work was more about construction of the project than exploring its feasibility and, in reading his open letter to Minister Chabot said, "How is the public and your ministry to get at the truth of the matter? Not, we submit, in a public meeting controlled by BC Hydro."

The degree of local concern was understandable. Notwithstanding financial implications, southern interests were planning wholesale alteration to a major ecosystem, the bottom line of which was also the bottom line in their regional economy where the livelihoods of many inhabitants centred on the river. If the perpetrators didn't know hard rock from soft rock, and didn't know if there might be gravel available, they should not be given responsibility for building dams anywhere, especially not across the Stikine River. Even should the hydroelectric project be abandoned, those two access roads from the highway would forever threaten ecological balance in the vicinity of the Stikine's Grand Canyon.

Subsequent public meetings in Dease Lake, Iskut, and Telegraph Creek were equally well attended and similarly vociferous, though exhibiting slightly improved levels of decorum. Continuing the dialogue, such as it was, the Tahltans (perhaps the Tribal Council) announced intentions of an injunction against Hydro if its appeal was granted. In a letter to the minister, the Sierra Club of Western Canada declared that granting of such permit in overriding the reasons for its previous rejection "would constitute a clear breach of the public trust and [a denial of] native rights." Months of pontification and recrimination finally came to an end several weeks later, in June '82, when the BC Ministry of Lands, Parks, and Housing denied the road-access appeal by BC Hydro, citing the same reasons that saw the rejection of Hydro's original application at the regional level. To many concerned citizens, it seemed saner heads had prevailed in at least one government agency. Suddenly, there was a glimmer of hope. A light at the end of the tunnel?

Alaskan Concern

Meanwhile, across the line in Alaska, concerns had been ramping up for some time. Within a year of it beginning studies on the river, BC Hydro was invited to Juneau by the US National Marine Fisheries Service to share its plans with state agencies. It was far too soon, of course, for Hydro to address projections on fisheries and other issues of importance to the people of Alaska. Nevertheless, it was time for Alaska to address the overall issue. Bringing together representatives from ten state and federal agencies, the Stikine–Iskut Instream Task Force was formed as the primary US contact with BC Hydro and as a conduit for coordinating information while volunteering to assist with further studies if requested. Soon thereafter, in March 1981, the Alaska Senate drafted a resolution calling on President Reagan to direct Secretary of State Alexander Haig to initiate talks with Canadian authorities to separate fact from fiction in regard to the hydroelectric development already underway upstream on the Stikine. By this time, Alaska's governor, Jay Hammond, was clearly on record as being opposed to any dams on the river.

The *Wrangell Sentinel* newspaper (March 25, 1981) reported the words of Governor Hammond from a press conference on the 17th: "I don't think it's an appropriate project. I don't think it's warranted. I don't think it's cost-effective. I don't think it's viable, and I think it probably should not be built from everything I've heard." While later explained-away as a "slip of the lip" misinterpretation of comments made by BC Premier Bill Bennett at their recent meeting, Hammond's strongest statement to date was probably a major factor, if not the only one, for weeks of foot-dragging and the rescheduling of previously agreed-upon meetings by federal representatives of both US and Canadian jurisdictions as well as by state and provincial officials. One week prior to meeting dates scheduled for early June—although the Canadian department responsible for external affairs had accepted its invitation—no reply had been received from Premier Bennett or any other Canadian agencies. Mail seems to have been the problem. Bennett's press secretary said that such a letter to the premier would have been handled by the inter-governmental relations ministry whose director of communications advised the letter had already been handed on to the energy minister whose assistant reported no such letter was received at the minister's office. Likewise, both BC Hydro and the DFO reported they had not yet received invitations. Dum-da-dum-dum.

Dialogue

After ten days of near silence on the matter, communication channels reopened with diplomatic notes between the Canadian embassy and the US State Department. It seems the Canadian federal and provincial governments, along with BC Hydro, had collective difficulty getting on the same page. A major stumbling block seems to have been the Alaskan organizers' plan to include a second day of meetings in Juneau to hear from interest groups, such as fishers, conservationists, and others for and against the dam project. With the Canadian contingent maintaining that each country should deal independently with interest groups, a one-day meeting was instead scheduled for Vancouver. Closed to the public and press, and hosted by BC Hydro, the gathering on June 30th successfully brought together the major players from both sides of the border to establish an arrangement for sharing information on an ongoing basis. Notwithstanding the expenditure of $30 million for feasibility studies, the Stikine–Iskut hydroelectric plan was clearly recognized to be a *potential project only* … and remaining subject to government approval.

"If the impact on the salmon is quite uncertain (as stated by you), how can it be termed 'minimal'?" asked Rosemary Fox of the editor of the *Vancouver Province* newspaper who had recently published an article suggesting BC Hydro's development would have minimal environmental effect. "Minimal is hardly the operative word for a reservoir 80 kilometres long, the level of which will fluctuate 150 feet! In no way can it be said that the impact of two large dams in one of Canada's recognized natural wonders would be 'minimal.' As you stated, the Grand Canyon of the Stikine is indeed an example of geological splendour." In closing, Rosemary pointed out that the concern felt by Alaskan authorities was sufficient to invite a meeting with Canadian federal and provincial officials, and the meeting postponements by Canadians "for considered response" were hardly indicative of a low-impact proposal.

Meantime, it seemed our public utility, BC Hydro, while already $6.7 billion in debt, was thinking it wise to overborrow six months in advance of need, subject to market conditions. A press clipping from the Business section of the *Vancouver Sun* dated April 7, 1982, captured the former Social Credit attorney general and now BC Hydro Chairman Robert Bonner decrying the meagre 11.5 per cent rate increase granted by the BC Utilities Commission as insufficient: "BC Hydro faces serious financial difficulty if it is unsuccessful in its bid for a 25 per cent rate increase … and with the stability of BC Hydro rides the industrial future of the province." He continued:

> *It would put a dent in our cash flow that would show up in our interest coverage, which possibly would show up in a review of our bond rating in due course … can make or break the credit rating, make or break our access to the market, make or break the industrial future of British Columbia. As one of the few North American utilities with a Triple-A credit rating, BC Hydro gets good interest rates and ready access to the markets. Failure to maintain the Triple-A rating would increase the cost for dams and other capital works projects, which is something we seek to avoid on behalf of our customers.*

While the big wheels kept on turning, albeit slowly, grunts on the ground were able to keep pace. Frustrated by the delays in cross-border "cooperation," Alaska's nascent Stikine–Iskut Instream Task Force continued to meet and identify priority areas for study while applying for budget enhancements of several hundred thousand dollars. Simultaneously, perhaps motivated by a visit from the chief administrator of the US Forest Service, Max Peterson, the Alaska Forestry Division took first steps toward a management plan for the recently created Stikine–LeConte Wilderness Area encompassing the entire US segment of the river as well as three major islands at its mouth (Sergief, Dry, and Farm). Peterson's visit was significant in being the first recorded by any senior official from either federal jurisdiction; moreover, accompanied by Alaska State colleagues, his multiday excursion included real time on the river and an overnight in a Forestry Division cabin as well as a weather-thwarted flight over the dam sites. "We're in the early stages of looking at the Stikine," he said. "That's why I am here."

On the Canadian side, no one was here looking. Deferring to corporate expertise and financial realities, the DFO allocated a grand total of $8,000 for study of the lower Stikine salmon fishery. Perhaps enough for two men and a boat for one month? Nevertheless, it was better than nothing considering the unknown effects of a hypothetical hydroelectric project backdropped by the eternal need for an international agreement on trans-boundary catches—aka the never-ending fish wars. Although protracted and intense as such wars usually are, the Stikine theatre seems to have extruded a degree of mutual respect between combatants as well as increased regard for the fish. Progress?

Fishing for Fish

Background: No Canadian commercial salmon fishery had existed on the Stikine before 1979 when BC Packers brought in a brine barge to collect the harvest from twenty-six gill net licences, a two-year exercise that proved financially unsatisfactory while netting 200,000 pounds (24,000 fish) in 1979 and 250,000 pounds (28,000 fish) in 1980.[6] With the departure of BC Packers, about a dozen Stikine fishers formed a cooperative known as Great Glacier Fishing (GGF) which had a 1,400-square-foot warehouse and a high-capacity freezer in operation on the lower river in time for the 1981 openings, all powered by a Pelton-wheel generator (with diesel backup). Other Stikeeners still had the option of delivering their catch directly to Wrangell, Alaska, or back upstream to Telegraph Creek.

The fishers of Wrangell were nevertheless unimpressed. Seen no longer as a small-subsistence fishery, the Canadian commercial venture was a sudden infringement on the take of US fishers who had, for ten years, been reducing their catch to rebuild stocks (according to them). Negotiations already in progress were aimed at ratifying an agreement before the 1983 season while permitting an interim status quo, providing the Canadian harvest did not exceed that of 1979, its first season in business. The Stikine run was being seen as one of the final points in twenty years of West Coast salmon interception talks aimed at allocating catches for both countries while ending boundary disputes by including allowance for future enhancement projects. Fortunately, so it was reported, the extreme highwater and excessive debris of 1981 disrupted the Canadian operation enough to prevent it from exceeding its prescribed sockeye limit. Nevertheless, dialogue between the two nations of fishers heated up dramatically over the definition of *traditional* and the interpretation of *limits,* or more specifically, "we were underfishing then, but deserving of a *percentage* share now, not a *number limit.*"

"Splitting fish runs is harder than splitting logs and more detailed than splitting hairs," wrote Larry Persily of the *Wrangell Sentinel* in reference to ongoing negotiations for allocation of the Stikine salmon fishery (April 14, 1982). While both sides recognized the imperative to maintain and improve existing stocks, the cross-border posturing increased dramatically: the Americans suggested they should fish as if the Canadians weren't there; the Canadians accused them of wanting to play Russian roulette until the stocks were eliminated. Was it okay for the Canadians to fish if they did not catch more than the Americans let them? Along with Canada's difficult-to-assess Indigenous subsistence rights came the higher political clout of an American fishery mostly dependent on larger and more expensive equipment needed offshore. Around and around and back and forth went various offers, attempting to find balance in good times and bad. "The stumbling block is not between fishermen and fishermen," declared GGF's Bob Gould, "it's between governments that cannot agree on management numbers."

The 1982 salmon fishing season on the Stikine began without any Canada–USA allotment agreement in place.

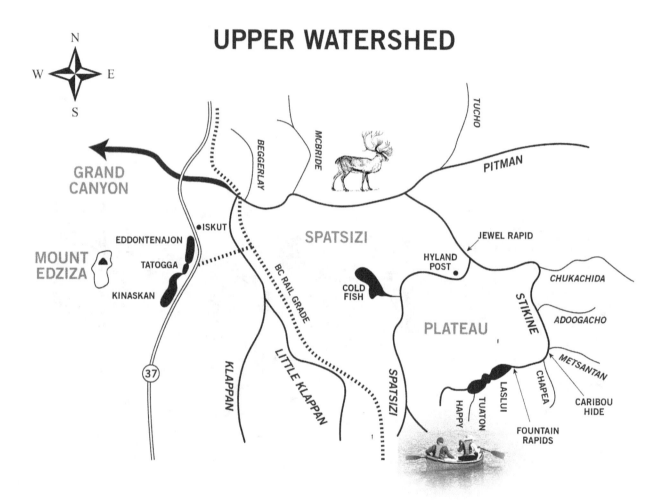

4
Leaping and Lunging

Meanwhile, life is what happens when you're planning something else. Long before seeing the truth in this adage, the reality of it hit me hard. Divorce. Though not particularly ugly, my separation from ten years of marriage came with the inevitable amount of pain and, eventually, a better understanding of life's complexities. Rapid progression from high school to airplanes to air force to air lines had been an exhilarating experience; however, there had existed little opportunity, or perhaps willingness, for introspection and self-definition. Having witnessed the ordeals of my parents who subsequently divorced, it now seemed far better to live happily apart than together in constant disagreement. Though my visit to Stikine country and my divorce occurred within the same time frame, they were not directly related. In fact, it was my about-to-be former spouse who had tipped me off about the initial Stikine River gathering at Robson Square. Thank you, Sharon.

Being back on "singles street" was an eye-opener. The bright lights of the big city soon illuminated the rough edges of a suburban cowboy who suddenly found the line between chivalrous courtesy and male domination to be extremely narrow and difficult to read. Although my attempts at being *swayve and duhboner* didn't measure up to James Bond, my fast-talking mouth got me into enough trouble. An art gallery pied-à-terre immediately outside the gates to UBC became my home, where I enjoyed new friendships among local artists and musicians.

Before long, I was a new student in the beginners' class for African drumming under the tutelage of Dido (later Mandido) Morris, a self-taught drummer who had escaped the ghetto of Watts in southern Los Angeles to eventually play and tour with the likes of Bruce Cockburn and Gino Vanelli. Watching and hearing him play brought stark reminder that, although all men are equal in the eyes of God, few have the gift of sound, and the grace of movement as did Mandido. He was an inspiration. Strong of character and impressive in physique, he was a great teacher, possessing extraordinary sensitivity and patience, along with a warm sense of humour. He says becoming a grandfather at age thirty-eight helped him stay in touch with humanity. Introducing African culture through the hand drum certainly helped me become better in touch with humanity, in the literal sense as well as the physical. His Sunday afternoon drum circles at Jericho Beach and at the old Soft Rock Café certainly did much for local humanity, bringing musicians and dancers together in lively community ... with this hockey player pounding along in typical fashion, high on energy and low on talent.

At one such gathering, a woman named Mary introduced me to the works of Bhagwan Shree Rajneesh—a mystic guru (teacher, mentor) from India who rejected institutional religions while encouraging human qualities generally suppressed by them, he prioritized meditation, mindfulness, and silence in a life filled with love, courage, and humour. The intrigue was compelling.

Indeed, the guru's words were not about religion or *a religion* but were about spirit and about being spiritual—life is a journey and love is the path—a notion that sat well with me. Church and religion had dropped out of my early life much the same as had music. Though the basic tenets of Christianity[7] remained with me, the power-tripping hypocrisies of church and politics had turned me off. Ultimately, it seemed to me, joining a particular church was the same as joining a particular political party: in both instances, obedience to an ancient dictum prevented appreciation of differing views in the modern age—ever-increasing partisanship hindered unity and forward progress. We seemed so attached to our particular ideology, we had forgotten we were all in the same boat: instead, we were so busy hyping our own schtick, we couldn't hear other voices that might be offering help and greater understanding. There must be a better way.

Life at Rancho Rajneesh, the guru's new ashram in Oregon, was eye-opening. Suffice it to say, there was more to it than was depicted in the media sensationalism. Yes, those who didn't understand the difference branded Rajneesh as the "sex guru" for his open acceptance of the sexual reality inherent in the human species. Of course, the realities of power politics and the frailties of human nature existed there as well, and they would eventually combine to undermine a courageous experiment; but in the meantime, there was peace in the valley and free transportation if you needed it. Here, a person's rugged journey into self-understanding was accompanied by a loving vibe in a supportive environment; meditation became a fact of life on the road to a meditative life ... at least, that was the plan. According to the masters, the road to wisdom begins with knowing you know nothing; and, of course, there is a large difference between knowing nothing and knowing you know nothing.

For some of us, Bhagwan soon became known as "Bugsy," in no small part because of his uncanny ability to provoke self-examination without saying a word. In short order, his meditation techniques had already rediscovered old injuries in my body while helping me know that *needy* and *greedy* were significant parts of my being. There was more to come. The highlight of any day at Rajneeshpuram ("Rajneeshville") was Satsang meditation at Buddha Hall—upwards of seven thousand humans seated together in cross-legged silence—amassing a powerful level of energy, difficult to describe but available to all. The sound of a cough, like a candy wrapper in a theatre, was startling in the extreme, forcing us further inside ourselves to find acceptance and an inner smile. Though Bugsy never spoke during this period, the reading of parables and quotations from him celebrated many masters, past and present, while encouraging religiousness before religion and spirit over dogma. At some point in some Satsang, it occurred to me that I was hearing something oddly familiar—something I had not heard spoken before and not being verbalized in the moment. Osmosis is probably an apt descriptor.

Though not remembered as an overwhelming factor, but a factor nonetheless, my parents' differing Protestant–Catholic faiths probably helped highlight the religious dichotomies in my small-town upbringing. Very early on, I realized all religions to be constructs of mankind (à la the Romans and Jesus Christ) contributing to the number and size of religious belief-*isms* permeating our society worldwide, all accompanied by the requisite number of schisms. For me and some others, going to church was poor substitute for a game of hockey or a walk in the woods.

Later in life, in not-so-rare moments of cynicism, I could be heard suggesting the Roman Empire to be alive and well and living in Las Vegas … while draining all the water out of the Colorado River. Yes, I knew human sanity existed on a plane somewhere between here and there … but where? Mass worship of a never seen "being" of many different names wasn't working so well for world peace and global hunger; instead, we appeared to be ravaging our planet while awaiting our trip to some *promised land* … which must be either another planet or a different state of mind. Assuming we are as greedy for peace as we are for almost everything else, maybe we should be looking closer to home. This world *is* our heaven and it's as holy as we make it. We are allowed to enjoy ourselves, but we must also take care of our campsite. We can be moneychangers if we wish, but we should never disrespect the temple. Perhaps, in their ancient scriptures, our early prophets were originally referring to *Heaven on Earth* and to *Love as our God*.

"No one said it would be easy." These words from my father have always been with me—a simple truth about the realities of life, especially when confronted by obstacles unexpected. Such philosophy of acceptance probably helped me gently push boundaries in numerous directions without demanding great success in any of them—a man of many interests and master of none. In the early 1980s, costume design became a matter of interest in the juggling of dark blue airline uniforms with the orangey reds of sannyasin wear. Oh yes, don't forget the string of meditation beads, all one hundred and eight. Being attired in distinctively different, outside-the-norm clothing was a challenge … like being naked; it was also a learning experience whereby I gained greater appreciation for newcomers to our culture seen wearing their traditional garb and/or having different coloured skin.

Of course, crossing the Canada–USA border southbound when dressed in red and wearing beads is a significantly different experience from crossing in a blue suit with golden stripes on the sleeves. Maybe the US border agents thought meditating sannyasins were retro hippies carrying drugs (no alcohol or drugs where we were going). Maybe they searched our vehicle, but they didn't search us. After sitting around and standing in line for more than an hour, we got the keys to our vehicle back and were told to be on our way.

Hmm. No harm, no foul, perhaps. But up came the questions about our society's professed open-mindedness and our so-called freedom of religion. Also now on the table were personal questions about the rationale for being different, for being in the spotlight, and for drawing attention to myself. Was it courage of being or just another case of being needy? Similarly, not understood at the time were the reasons why I was ineffective as a parent, part-time or otherwise: neither marriage nor child-rearing had been subjects of education offered at home or in school. I was nevertheless fortunate to be sharing time with my children in parts of my life as well as in parts of theirs—a tentative loving bond in tumultuous times. "Selfishness" was still an unacknowledged part of my being.

Meanwhile, at my new bachelor pad, Gallery ANX, near the west end of Tenth Avenue in Vancouver, coloured photographs of the Stikine River were flowing freely over whitened walls—the idea of making a motion picture was already in bloom. A motion-picture capturing Stikine's wild beauty seemed an easy-to-make tool for helping celebrate the river's natural state, and the possibility of *failure* had not crossed my mind. If people saw the river and saw what was at stake, they too would prefer it as a natural resource rather than a manufactured resource. I was not an environmentalist then and I would never come to consider myself as one. From the get-go, I was simply one concerned citizen joining other concerned citizens in a dialogue around the value of free-flowing rivers and natural ecosystems. Not a true cause célèbre, the Stikine was simply front and centre on the path of a new believer—Earth is our heaven and love is our God.

Back to the Bush

One fine evening in Vancouver's wild West End, friends introduced me to Gary Fiegehen, a professional photographer who had just moved into their building where we soon became wrapped in conversation about my favourite subject outside of airplanes. Losing interest in shooting oil refineries and annual report items, Gary became extremely interested in knowing more about the Stikine River that, he claims, had me in a fever pitch. In harmony with my photographic wannabe status, we soon found common ground on the subject river and became committed to going there together sometime. And it did not take long for the bush telegraph to identify others wishing to join us: Monty Bassett and Carl Chaplin were both known to me through mutual friends. Writer Monty and photographer Gary immediately connected through *Equinox* magazine to do an article about the Stikine. It did not take long for the idea of a Stikine River canoe trip to launch itself and we began putting the necessary elements together. A former rodeo clown living on his ranch near Smithers, Monty was executive director of the Spatsizi Association for Biological Research, which currently had him leaping out of helicopters to radio tag Osborn caribou on Spatsizi Plateau—and he had a canoe. A lifetime creator of extreme visual imagery currently living in an abandoned ammunition warehouse across the tracks from an abandoned WWII airstrip near Kitwanga, Carl was looking for new things to paint … and yes, he said he had been in a canoe. A Ryerson-trained photographer and sometime theatre manager now living in the West End's delightful Capistrano building, Gary was an inspired taker of pictures looking for a return to the land … and yes, he had paddled a canoe. An airline pilot with Stikine River fever, Peter's previous experience up there seemed all that was necessary … and yes, he had a canoe … and a truck that would carry two canoes.

For Gary and me, on a double date with our female partners, the summer of '82 began with a two-day trail ride under the guidance of Willie Williams who introduced us to the Grand Canyon of the Stikine by horseback. One look at Entry Falls made the entire trip from Vancouver worthwhile: no signs, no fences, no guard rails—just raging water beneath towering walls, an immersion in immensity. Riding the rim and glassing for goats was a warmly satisfying occupation. The sight of Willie standing on a far-removed skinny point of rock with binoculars to his eyes—being comfortable in one's habitat—underlined my own fear of heights. Barely surviving a sudden canyon-top sprawl to save a wayward camera lens also brought me in touch with my own *stupidity,* bravado notwithstanding. All in all, it was a wondrous introduction to the canyon and to its people. With departure of our dates toward southern climes, we boys headed for the hills—two (fool)hardy adventurers on the move.

Enroute to our destination in the southern part of Spatsizi Plateau with two aluminum canoes strapped to the struts of a sturdy DHC-2 Beaver, it was an aviator's pleasure to observe the measured calmness with which pilot Ron Bruns adjusted throttle settings to his big round engine—just enough power to clear the high ground without pushing his engine temperatures through the red line and us into the cloud above. Sweet. Though our boats were relatively light, they would have added a half-ton of profile drag. (Regulations for external carriage of canoes on aircraft have reportedly been since amended in the direction of increased safety—only one canoe is now allowed.)

After stowing our boats at Tuaton Lake, Ron graciously hopped the two of us and our camping gear up to Happy Lake for more convenient access to the higher terrain I had experienced with Hal in '79. This time, the weather was shitty, especially so for a dedicated photographer like Gary. A solitary caribou ghosting through the mist and mediocre photo shots from atop the high waterfall were poor compensation for a two-thousand-foot climb followed by a tent-bound day of heavy rain. Our hike back to Tuaton proved much more than expected: our straight-line "shortcut" gave us skinny beaver dams to tightrope and cold streams to wade with our packs held high. Live and learn. My filmmaking project was suddenly more difficult than imagined.

One day behind schedule, we arrived at the mouth of Happy Creek to connect with Monty and Carl. They'd flown in directly from Smithers and were trying to remain patient while camped out beside our canoes. After regrouping and reorganizing, an incident-free paddle over several days in mediocre weather got us to Sanabar Creek at the northeast corner of Spatsizi Plateau where we became drenched in a heavy night of rain that seemed in no hurry to move on … much like us. Slightly downstream of the Spatsizi confluence, with the rain easing off, we decided to exchange paddle partners for the first time … all in a spirit of congeniality to brighten up the day.

My vast experience of having been on the river once before had me in the stern seat of the lead canoe, ostensibly guiding three friends who happened to share deep concern for this river, which none of them had ever been on. A convivial getting-acquainted conversation with new bow-mate, Carl, was in high gear when we suddenly found ourselves confronted by a noisy rush of white waves in a confusing array of hard-to-see boulders. Welcome to Jewel Canyon! Though not seen as immediately perilous, the rocky kilometre ahead looked more challenging than expected. Oops!

A quick paddle brace after a keel-rub on a hard pillow saw us safely down a narrow chute … no worries, Monty's an experienced paddler and the route is optional. By the time second thoughts arose (the other canoeists were probably also lost in conversation while blindly following us) the current had carried us too far downstream to issue any effective warning. Hand signals would be useless because none had been established (ugh!), and barely audible voice instructions from a distance could have been confusing for those about to enter the narrow chute. Pointy end first was best. All we could do was drift backwards and watch.

Over they went in near unison, both fortunately wearing personal flotation devices. As Gary and Monty swam their boat towards river-right, we hovered downstream in mid-current, on the lookout for escaped articles. Sure enough, after deciding we could delay no longer, a paddle and a map disappeared with the current off our stern. Ouch! A large, cozy fire on a cold, rocky beach was slim comfort for two fatigued bodies dragging ashore from a five-minute immersion in lethally cold water. The need to physically force our tormented souls into warm sleeping bags made the distracting effects of hypothermic reaction readily apparent.

This was a big wake-up call for me. Though significantly more challenging than my previous low-water experience there and fortunately far less threatening than later seen, Jewel had enough kick at this level to unseat the unprepared. Although others in our two-canoe party had considerable backcountry experience, they saw me as the de facto leader of the group by dint of my previous experience on the river. While my aviation career made me familiar with responsibility, the factors of leadership and guidance were yet undeveloped qualities waiting to emerge from my growing collection of mixed experiences—detectable at times, but never established, they had certainly not appeared in my new world of wild rivers. Though comfortable in the role of default leader for overall direction of this adventure, the specific role of river guide was not in my skill set. My moving-water credentials were appalling and my experience equally so. My initial attempts at book-learned technique had once dumped my young family into the Chilliwack River, and only by hands-on teaching of elementary bow-strokes did Hal get me down this river … once.

Although I was naively confident of finding my own way down, the idea of telling others where to direct their boats had not crossed my mind. Nevertheless, a boatload of regret would always accompany this "sleeping at the switch" learning moment. It would be another ten years before the real joy of white-water paddling (and its necessary discipline) came to me through the teachings of Tony Shaw and like masters. Meantime, after warming up the bodies and packing up the gear, we slipped out the bottom of Jewel to enter a lengthy stretch of homogenous landscape on a current flowing with ever-increasing momentum. Any doubts about the river being high and rising were dispelled overnight when our shallow beach campsite was threatened by an overnight rise in river level. Not only was the river significantly higher than on my first visit, but it was now getting higher by the day or, more correctly, by the minute. The vast landscape above and behind was collecting rainfall on a recently regular basis and draining it into scores of streams feeding a dozen tributaries funnelling into our solitary channel, which was now squeezing west at a healthy clip.

After rounding another nondescript bend in the river, suddenly there it was, Beggerlay Canyon. WOW! A classic example of power and beauty out here in the middle of nowhere. I had mostly noticed its beauty on first meeting. This time, there was great power in sight and sound … being pushed toward rocky sentinels guarding a narrow opening with a foaming ledge across its entrance … slicing into a large eddy on river-right to make shore on a steep rise of boulders, just short of

Beggerlay Creek which crashed in with a mountainside of fury. The creek is loud, the canyon is intimidating, the place is amazing. Hauling out over the steep and difficult footing, I marvelled at the dynamics of the place. How do thousands of basketball-sized boulders get pushed into a wall of "river rocks"—in a back-eddy? Up top, about six feet above the water, a flood-plain bench of level sand provided camping convenience, and another step upwards into the trees offered comfortable bedding. The raging creek never sleeps.

Immediately upstream of the creek mouth, an underwater ledge extends almost halfway across the river, creating a wide swath of uninviting territory. A short, narrow chute alongside a rock wall on river-left is the only way past the creek-mouth maelstrom and into the canyon proper. The current scene was intimidating, far more than anticipated. Perhaps Monty's cold soak at Jewel had coloured his look at Beggerlay with a darker lens. In any event, he and his partner Carl decided to portage their canoe and gear past the threatening area at the canyon entrance. A sensible idea … except, the only way *around* the perceived danger was *across* Beggerlay Creek. No easy feat. From experience, I knew that wading knee-deep across high-country tributaries was possible by respecting the power; being waist deep in this violent rush of mountain-fresh water cascading down a sloping bed of slippery boulders was a different matter. Monty and Carl did prove it possible, but only with the aid of ropes and belays, along with the assistance of their companions. This very short portage was a life-threatening ordeal of several hours in duration. The sight of another paddle being swept to oblivion put an exclamation mark on the danger sign.

Thus, it was easy to understand my partner Gary's willingness to go along with my Plan A, which had been developed through the study of manuals and videotapes from the genius of Bill Mason about introductory white-water techniques. Surely, my plan could not be any more dangerous than "fording" Beggerlay Creek. With help from the back-eddy current, we paddled like hell upstream as far as possible and then sliced into the main current (without tipping over), then paddled like hell toward the far shore until death and destruction seemed imminent, then turned quickly downstream and entered the chute, pointy end first. It worked. Our paddle strokes were more panicked than polished, but enthusiasm prevailed for the smooth rush of a short, steep slide into the running bounce of mid-canyon rollers. Yee haw!

Happily, unscathed, and unsaturated, we four river rats, having gained some experience, passed beneath the never tracked $3 million railway bridge of 1972 while simultaneously acknowledging the surge of big Klappan River coming in from the south with its host of tributaries from the western side of Spatsizi Plateau. With this added boost in volume, Stikine soon upped our attention with Little Beggerlay, the unofficial name for an innocuous looking riverbend on the map, which produces another hit of river awareness just before the finish line. It's never too soon to learn and never too soon to apply a paddle brace. Pulling out at the highway bridge, four (fool)hardy adventurers drove to Telegraph Creek in my trusty Landcruiser, which now sported four nearly new tires and carried two never used spares, on the long road to wisdom. While pausing in appreciation of the

Stikine–Tahltan confluence, a photographic selfie identified the adventurers: from left to right, Carl Chaplin, Monty Bassett, Peter Rowlands, and Gary Fiegehen.

Telegraph Road was as rough and beautiful as before but deflated no tires. Rustically charming, the village had received its start and its name by being a staging point on the abandoned Collins Overland (Western Union) telegraph line of 1866 before becoming the Stikine River crossing site for the BC–Yukon telegraph line of 1901. With time to spare (no pun) in 1982, there was ample opportunity to dawdle on the waterfront, learning more about the place and its people. Francis Gleason, who had pointed me toward tire rescue three years earlier, proved to be a long-established fixture on the lower Stikine, providing reliable riverboat service anywhere between town and tidewater for anyone in need.

At the other end of the street, the large RiverSong building was a former Hudson's Bay Company trading post, originally located twenty kilometres downstream at Glenora Flat, once a junction on one of several gold-rush trails to the Klondike. (In the winter of 1903–04, the store was cut in half and moved by horses upstream to its present site, where it was put back together to serve as a trading post and general store until its closing in 1972.) Happily, on this day, we discovered it was again open for business, operating as the RiverSong Café and General Store, a multi-family cooperative also offering accommodations and riverboat services under guidance of its principal owner–operators, David Fisher, and Dan Pakula, with whom we quickly became acquainted.

Also, in the store that day was a visitor from Iskut village, located on Highway 37. Having helped stir up BC Hydro awareness in Wrangell, Alaska, Jim Bourquin[8] was also a river guide like Dan Pakula, offering raft trips in the lower canyon. Jim knew the entire river: one photograph showed him running a raft full of people down the rocky boils of Fountain Rapid in the summer of '79, about the time of my first visit. Impressive.

Residents

Being in the company of local river rats, it was easy to segue my ignorance into questions about their group, Residents for a Free-Flowing Stikine, and to query the latest news. "Come with me," was Dan's simple directive, leading us down a row of outbuildings to enter a warehouse and find people and paperwork in action. Welcome to Residents!

On duty were Cheryl Reitz and Phil Lander, both of whom were knowledgeable and happy to help. When not assisting in the store, restaurant, or lodge, Cheryl coordinated paperwork and activities for RFFS, which she points out was started by the RiverSong Guys (not Gang)—Tony Bute, Dan Pakula, John Plummer, and Joe Murphy, with of course, significant help and understanding by their respective others, Rachel, Diane, Rosemary, and Nancy—all from among numerous homesteaders strung out along the Glenora Road west of town. Difficult as it was getting any number of free-thinking "pioneers" to agree on anything, their major concern was landlords—the Tahltan peoples—on whose lands they had come to settle.

Without dwelling on the rationale or legalities of the recent past that allowed them to homestead on non-treaty land, the residents of the day were challenged with the need to respect the will of their Indigenous neighbours, while being true to their own strong belief in the river's preferred state. As in any community, especially when poverty exists, strong arguments prevail on both sides of the progress debate. Communications beyond the community were difficult. Choose one: an expensive, scratchy, party-line radio phone or a month-long wait for return mail. Operating costs were being eased by a small community-aid grant courtesy of Jim Fulton, their well-respected member of parliament.

As a summertime volunteer from away, Phil was able to offer some overview of recent happenings. Apparently, this year's annual heads of state meeting held in Juneau (Alaska Governor Hammond, BC Premier Bennett, and Yukon Premier Pearson) included an agreement between Hammond and Bennett for information sharing about the BC Hydro project, and a cross-border committee had been set up to deal with it (Alaska–B.C. Information Exchange Committee). Alaskans were concerned primarily about effects on the fishery, their primary economic engine, which is dependent on Canadian spawning grounds: they worried that Hydro would get the go-ahead on construction before their US studies were completed. Rumour had it that Bennett didn't think any application would be made before 1984, after completion of studies in Canada.

There was no resolution to the fish wars yet. Canada was talking about trading coho for sockeye. "Bob [Gould] and Stefan [Jacob] down at Great Glacier [Fishing] have added an ice-making machine this year so they won't have to collect from the glacier … they now have a Japanese guy on site processing roe for sale back in Japan. Some SEACC people [Southeast Alaska Conservation Council] will be here in September … they're planning a raft trip to raise money. Should be fun."

Back in the main RiverSong building, the lodgings were comfortably rustic, and the home cooking was superb: good enough, we learned, to lure some regular users of Highway 37 into 250-kilometre round trips for Diane's fresh blueberry pies. Tasty, indeed. When Monty returned from the nursing station with a diagnosis of mild pneumonia, probably promoted by his earlier swim upriver, he was advised to temporarily remain off the rain-soaked river. We had to modify our plan. As such, Gary and Peter continued downriver on their own to Great Glacier campsite to reconnect with Monty and Carl several days later when they arrived with David and Dan on their riverboat. Having no immediate openings for fishing, the locals strapped canoe *Dimples* to the top of their Bimini and took us tourists into Ketili Slough for the ultimate lower-river treat—a soak in the tubs at Chief Shakes Hot Springs. Too good to miss and (almost) too delightful to describe: two large tubs, one indoors with mosquito screens and one outdoors with fabulous view; two large-diameter black hoses from different creeks, one hot and one cold; mix and match as desired. Yee haw!

From there, less than an hour of powering around, over, and through the delta's maze of ever-shifting sandbars got us to Wrangell, Alaska, for US Customs and Immigration. As Dan and Dave (our new best friends) headed back to their duties on the river, we four cheechakos indulged in the obligatory photo op before imbibing a celebratory pint or two. We had been lucky. Nevertheless, leaving Stikine is never easy. Gary stayed in Wrangell awaiting better weather and the arrival of Ron Bruns and his Beaver for an upstream photographic mission to Telegraph Creek. Along with my canoe, we three others hopped the Alaska State Ferry to Prince Rupert, from where an airplane took me back to duties in Vancouver while Monty and Carl took *Dimples* with them on the train to Smithers.

Almost a week went by before Gary telephoned: the weather hadn't been great when Ron finally got them airborne, but he thought he had some good shots; unfortunately, on the way out, he had gotten into some loose gravel on the Telegraph Road … the Landcruiser had rolled onto its side and slid part way down an embankment. Monty's canoe was not damaged, but Brutus had to be towed to Smithers for repair; Gary and Brutus were then on their way to Vancouver with *Dimples* now on top. Fingers were crossed for good luck. Two trips on the Telegraph Road and two memorable stories written in the mechanical genre. There would be more. Meanwhile, one of Gary's first aerial photos out of Wrangell was a look upriver over Sergief Island, a classic image destined to become a poster and a book cover.

Leaping and Lunging

5
Chattering and Clattering

The transboundary fishing debate of 1982 took on a lower level of intensity as the season neared its closing date. While fish counts were being undertaken to determine the level of compliance with a non-existent policy, biologists from the Alaska Department of Fish and Game were on shore and in the air to study possible effects of BC Hydro's project on the lower Stikine moose populations. Based on projections from BC Hydro, water levels near the river's mouth would drop approximately 0.82 metres in summer months and gradually, in fifty to one hundred years, invading stands of spruce and cottonwood would replace some of the red-osier dogwood and willow. While the highly adaptable moose were considered unlikely to go hungry while losing their favourite browse, the hunting fraternity of the southeast would find some of their favourite predator sloughs inaccessible because of the lower water level.

Considered to be stable in numbers and well nourished, the Stikine over-wintering herd of 1982 consisted of approximately 175 animals with thirty-eight calves and five bulls for every one hundred cows; the low bull-to-cow ratio was largely attributed to heavy harvest by hunters who were each allowed one bull. In late October, the *Wrangell Sentinel* reported thirty-one bulls were taken in the thirty-day season ending on October 15th.

BC Hydro neglected to comment on the matter. Perhaps of little relevance to the southern planners of power, this glimpse at the lower Stikine's moose population opened my eyes to the potential effects a remote man-made structure might have on the flora and fauna of a given bioregion. While being introduced to the area's systemic complexity, I began to wonder how many other species and habitats were eligible for consideration. Though not a major concern at the time, especially for a remote seagoing estuary, a 0.82 metre difference in water level certainly has the attention of today's coastline inhabitants.

First announced in August, BC Hydro's plans for public meetings in 1982 at Wrangell, Petersburg, and Juneau to review its Stikine–Iskut hydroelectric project had been set back a month, from October to November. Apparently, according to the same newspaper that reported on the lower-river moose bulls, budget cutbacks at the provincial utility had caused the delay. "The economic recession is going very, very deep into BC Hydro, forcing it to lay off workers and to delay many of its projects," reported Eric Powell of BC Hydro's community relations office in Vancouver while expressing a desire to distribute a Stikine–Iskut environmental study report in time for the meetings. Contrary to prevailing skepticism, BC Hydro did approve and print its study results—an eleven-page document giving interim conclusions for three or four years of study—made available, free of charge, prior to public meetings in Wrangell and Petersburg. The idea of a public meeting in Juneau was dropped in light of a forthcoming second meeting of the Alaska–B.C. Information Exchange Committee, which was scheduled for there and open to public participation.

Meanwhile, in Petersburg on October 20th, a meeting of the state–federal Stikine–Iskut Instream Task Force brought together biologists, hydrologists, and other "ists" to review data gathered from summer studies on the lower river. Also present was Ken Wilson of Beak Consultants, a Canadian firm retained by Hydro to study the lower river. "In three years on the job, the most common question I get always refers to the fear of any change in access to Shakes Hot Springs," Wilson remarked in good spirit, "but the more important questions, from my point of view, are from the US specialists. *These are the people we can't fool.*" [Italics mine.]

Comparing themselves to chefs of different restaurants preparing for a banquet, this assembly of scientists was unanimous in its appreciation for the opportunity of being together and comparing notes wherein sediment load was a big factor for almost everyone. BC Hydro hydrologist David Jones and his team, who collected 1,500 suspended sediment samples from eight locations, predicted the dams on the Stikine would trap two million tons of sediment *per year* and those on the Iskut would retain an additional 3.5 million tons *per year*. While believing there would be a reduction in growth rate of the estuary with dams in place, Jones did not foresee any substantial reduction in riverine vegetation during the short term. "One of the things we don't have a shortage of on the Stikine is sand; it's awesome," added Wilson. "A 35% reduction in estuary growth would hardly be noticed: at the rate it's going, as in coming downriver, the idea of some day being able to walk from Wrangell to Petersburg is not too far fetched. More importantly, we are concerned about non-visual changes in the delta because the river mixes water on the edge of the delta into the seawater and adds nutrients to the seawater … a change in the nutrient level could lead to a change in the plankton community—the bottom of the food chain. Ultimately, we are concerned with fish … we don't eat habitats."

That meeting's background notes provided some oft-forgotten numbers, which are quoted here in their natural non-metric state, along with today's newly created formula for measuring such things: the Site Zed (Canadian pronunciation) dam would have a height of 890 feet and a generating capacity of 915 megawatts; the Tanzilla dam, slightly downriver, would have a height of 663 feet and a generating capacity of 915 megawatts; the Iskut Canyon dam would have a height of 518 feet and a generating capacity of 780 megawatts; the More Creek dam, on an Iskut tributary, would have a height of 442 feet and a generating capacity of 155 megawatts; the dam on Forrest Kerr Creek would have a height of 121 feet while diverting its flow into the headwaters of More Creek. Good stuff: 2,634 vertical feet of dam (half mile) produces 2,765 megawatts of electricity; that's good mileage on high-octane water—slightly better than one megawatt per foot.

As anticipated, the level of public participation at the subsequent November meeting of the Alaska–B.C. Information Exchange Committee in Juneau was substantial. Mary Ellen Cuthbertson, speaking on behalf of the Southeast Alaska Conservation Council, stressed the need for studies to examine the effects of the hydroelectric project on the Tahltan peoples who live there. Similarly, the Telkwa Foundation, an independent research group based in Smithers, BC, expressed concern over the lack of social and economic analysis of the project's effects on residents of the Stikine Basin as well as on those peoples to the south through whose lands the transmission lines would be routed.

Allan Stein, a long-time American commercial fisher, was emphatic in stating the proposed dams would have disastrous effects on the commercial fisheries of Wrangell and Petersburg: "Who's going to pay for the loss of livelihood of fishers, mechanics, and storekeepers when the fish runs are lost? Monetary compensation for lost business and aquaculture programs to offset the loss of natural salmon runs should be figured into the overall cost of the project." A letter received from Discovery Alaska also pointed out the substantial contributions made by guiding companies to the area's economy would be lost if dams were built.

Responding to the public testimony, Everett Kissinger of the committee's technical task force (and Stikine area manager for Alaska Forest Service) acknowledged the need for involvement of other state agencies beyond the technical forum in order to study social and economic effects of the project; such involvement would be made easier if it became a priority of the incoming administration. (Bill Sheffield had recently become Alaska State Governor, replacing Jay Hammond.) Chuck Harmon, the non-voting appointee on the BC panel and BC Hydro's project manager for the Stikine–Iskut development, reported the utility had for the moment dropped plans for building access roads and airstrips in the Stikine as well as cutting back activities at base camps doing exploratory work—if BC Hydro's application was approved, most recent projections showed construction to begin in 1988 and power coming online in 1994.

In seeking an early meeting with BC Premier Bennett and Yukon Premier Pearson, Alaska's new governor, Bill Sheffield, advised he would be more development-minded than his predecessor (Hammond) and he wouldn't oppose BC Hydro's dams as long as they did not damage Alaska fish stocks. When reminded of his campaign rhetoric which suggested "… progressive and aggressive action by the state's governor could halt development of the Stikine dams," he suggested the possibility of being misquoted back then. "We need power and we're building hydro projects in Alaska, too," stated Sheffield, "I have not yet had a chance to study BC Hydro's proposal in detail, but, if we can work things out so that our fish in Alaska are safe, then I would be satisfied with the project."

Meanwhile, in downtown Wrangell, the local chamber of commerce was passing a resolution in support of BC Hydro's Stikine–Iskut power project. In presenting his motion, council member Bob Urata stated, "I know it's going to be controversial, but the project would stimulate some commerce in Wrangell and enhance road development." (Urata and others in the community had long advocated for a back-channel connection to the interior highway system.) "To oppose it doesn't give us any leverage at all," suggested Lloyd Harding; "We should go with it and ask for concessions from BC Hydro, such as money for fisheries enhancement on the Stikine to rebuild salmon runs." However, as reported by Larry Persily of the *Sentinel*, "Though many at Thursday's chamber meeting talked of needing more information on BC Hydro's project, few chamber members attended BC Hydro's Friday night public information meeting at city hall."

Subsequent letters to the editor of that newspaper suggested some dissension existed within the community. One such letter from Joseph Sebastian warrants quotation: "… this proves the chamber is willing to put the tinkle of coins in their pockets ahead of a free and mighty river … the very soul of this country. Perhaps if we crawl on our hands and knees and 'go with it and ask for concessions' we will be granted a lousy little fish hatchery to replace the strong natural runs that used to live in the river for thousands of years before the river was dammed … can we stand idly by while the Stikine is sold down the river and our heritage gambled away by anxious local and state officials with visions of sugar plums and dollar signs dancing around in their heads?"

Similar sentiment was also found on the editorial page of the *Alaska Fisherman's Journal*: "… It's hard to get worked up about a deal like the Stikine dams now, though. There's no shortage of hearings and meetings; the EPA is requiring an EIS on the MBI; and the friends of this are fighting the friends of that. You have to be a full-time pro to figure out what's happening at all. Of course, if a guy wants to build a controversial dam he won't need for another 20 years, he'll start now, slowly, and quietly, and speak nothing but Lithuanian so nobody on the opposition side can figure out what's going on. … Multiple resource is a good idea, and a fair one. But fishermen don't have to be unbiased philosophers at meetings or in print when they're talking about dams and strip mines. Leave conciliation to the mediators and hold fast to the position that one salmon lost to a dam or strip mine is one too many."

Out in the real world, wildlife biologists Dan Rosenberg and Patricia Heglund were adding up their numbers after six months on the Stikine River delta doing research for the U.S. Fish and Wildlife Service. Although the year's observations turned up no real surprises, it did produce items of interest. About 25,000 snow geese (the largest number yet recorded) had made a pit stop on the flats for food and rest during their southbound journey in the fall; similar numbers also stopped there in the spring season while travelling north to their breeding grounds on the other Wrangell Island (Wrangel), off the northeast coast of Siberia. Tracking information suggests most of these birds spend winters in Puget Sound north of Seattle and/or at mouth of the Fraser River near Vancouver (close-in on final approach to runway 08R at YVR).

The most abundant types of ducks seen during shoulder season travels were mallards, pintails, widgeons, and green-winged teals, with numbers traditionally rising and falling in accordance with moisture content of the interior plains. As well as Canada geese, white-fronted geese, and trumpeter swans, thirteen thousand sandhill cranes transited the area, two thousand of them stopping on Sergief Island. Though difficult to quantify, fifty thousand to one hundred thousand western sandpipers are deemed to have utilized the shoreline within the year. The yellow warbler was the most abundant among thirty-two species of dicky birds encountered. As expected, the spring run of eulachons (known locally as hooligans and elsewhere on the coast as oolichans or candlefish) attracted a significant number of bald eagles—482 birds and seventy nests. (The highest total ever recorded there is 1,500 bald eagles on the river at the same time.)

"BC Hydro has spent about $40 million on engineering and environmental studies … and the State of Alaska has appropriated almost $1 million for fish and wildlife studies … but no one has looked into what effects the $12.6 billion project might have on Wrangell's economy or its residents." So began a *Wrangell Sentinel* piece by Larry Persily (March 9, 1983). Although agreeing on the importance of such a study, the Alaska–B.C. Information Exchange Committee could not decide which party (Alaska or BC) should be responsible for its completion. When asked if his company had plans to do any such study, BC Hydro project manager Chuck Harman said none were planned because no one has asked them to do any. Don't shoot the messenger.

In a *Vancouver Sun* interview of June 8, 1983, Chairman Robert Bonner explained how the current economic slump was delaying BC Hydro's plans. "Faced with too much electricity chasing too little demand, Hydro is delaying its next major projects by at least two to three years … and is considering changing the order of its next generating facilities to bring new power on-line in smaller increases," stated Bonner. Acknowledging electrical output in BC to have been flat over the past three years and contrary to previous estimates of ever-increasing need he had strong recommendations for righting the good ship Hydropop, which had reduced its staff by fourteen per cent from its 1982 peak.

BC Hydro now wanted permission to sell electricity to the US on a "firm contract" basis—guaranteed long-term—while interest rates were "monkeyed down" by the central banks to improve borrowing power. Mr. Bonner also stated: "The objectives of the 'no-growth' advocates have been achieved since 1980 and I do not think anybody is applauding the result … though exports of electricity to the U.S. has helped keep Hydro profitable over the last three years, they cannot be relied upon every year … no, we can't afford to help homeowners conserve energy … BC's major electricity consumers such as pulp mills and saw mills have already been very successful in conserving energy." From today's perspective, it seems Mr. Bonner and company had priorities other than providing power to the people of British Columbia.

Uh-oh! Coincidental to this apparent retreat by BC Hydro came first clues about other "invasive species" in the Stikine. Re-reading my winter of 1983/84 newsletter, referencing the first timber sale on the Stikine in twenty years and news of an impending gold mine, it is obvious that I had become infected with a relatively serious case of *polarization*. Although my "invasive species" term had flowed easily onto paper in lighthearted context as a form of begrudging respect for the "other" side, it was in fact a case of the pot calling the kettle black, a case of the tourist calling industry a visitor. Local residents seldom consider an industry invasive when it offers paid employment. While the image of artificially enhanced salmon swimming upstream forces a closer look at the terms "invasive" and "natural," it does little but lead the discussion straight back to nowhere. Yes, all humans are invasive in some way and big-time consumers in other ways; however, in the glow of self-innocence, we small-time paddlers see ourselves as but bits of bacterium in the bloodstream of God, drifting righteously along between toxic mines and cancerous dams—we are not the disease, just the germs.

Interestingly, the BC Ministry of Forests—which chose not to attend Sierra Club's initial 1980 workshop and which had since maintained the watershed contained little timber of interest—held a timber-sale auction on October 3, 1983, at its office in Dease Lake, where the only bidder present,

HAL–PAC Forest Products of Kitwanga, BC, paid the $24,000 minimum price for harvest rights to 60,000 cubic metres of cottonwood and Sitka spruce on approximately 500 hectares at the Stikine–Iskut confluence. Ian Bowie of the forest ministry said this unusual sale was a pilot project to determine economic feasibility. HAL–PAC President Ray Halvorson (who probably initiated the original idea) estimated more than ten million board feet of harvested cottonwood would be rafted to Wrangell and loaded onto ocean freighters bound for China to become chopsticks and packing boxes. In very short order, a double-decker 24- by 48-foot steel barge was tied to the river-left shoreline at the Stikine–Iskut confluence where it housed a fifteen-member logging crew together with office and workshops. By mid-November, the trees were coming down.

About forty kilometres upstream on the Iskut, still on river-left, Johnny Mountain provided Reg Davis, president of Vancouver-based Skyline Explorations, with encouraging news. After three years of concentrated work in the area, with as many as thirty men, came extremely promising results and reasonable certainty of a gold mine being established, possibly also in conjunction with healthy lead, copper, and zinc deposits if warranted by base metal markets. Meantime, gold prices were attractive and stable, such that a mine seemed inevitable. With it came the need for access. So far, the exploration camp had been supplied from Wrangell by boat to Johnson's Landing, slightly upstream from the Stikine, and then shuttled the last forty kilometres by helicopter.

An operating mine would be another matter. Already being contemplated was a forty-kilometre road from the mine down to Johnson's or a fifty-kilometre road up the mine's adjacent Craig River to connect with Bradfield Canal on the coast. Either option would be expensive to implement and horrendously difficult to maintain, especially in winter over the Coast Range divide. To many of us, both these road options looked highly undesirable … irreparable scars inviting more trouble. Some roads are more welcome than others.

In those days, the southern four hundred kilometres of Highway 37 offered three villages, three gas stations, and very few accommodation options. Unexpectedly, a particular roadside attraction ten kilometres south of the Eddontenajon float-plane base would become my home away from home during numerous subsequent visits. Located on the highway and near the centre of the watershed, Tatogga Lake Resort featured good food and friendly atmosphere, along with equally important gas pumps and tire repair for highway travellers. Furthermore, by having cabins on the lake with a float-plane dock, it was an ideal base camp for two mostly airline pilots who had assigned themselves to aerial photography missions over the Stikine during the summers of '83 and '84. A long-time mentor on the airline and on the ski hill, Bill Murray had great respect for BC District and for wild places in general, such that he flew the two of us north in his family's float-equipped Cessna 180, CF-VSB (Very Special Bird) to photograph the entire river from top to bottom and back again. The fact that Mike Jones, our amiable host at Tatogga Lake Resort, was himself an experienced pilot ensured no lack of lively conversation on many subjects. The fact that Mike was also a fisheries biologist who had consulted on BC Hydro's initial studies for the proposed dams helped him provide realistic perspective on the Stikine—and then some.

Launching his aluminum river boat at the highway bridge, Mike took Bill and me where few people have ever gone, and he gave us a look at the river we would otherwise never have had. In warm, sunny weather, the scene bordered on surreal: smooth-gliding S-turns between towering walls along a mighty river stampeding straight west into the Grand Canyon of the Stikine. We didn't launch off the top of the infamous Entry Falls, but Mike took us as far as possible without killing our story. Pointing upstream with a handful of throttles, he backed us right up to the slippery lip of the falls where—immediately beyond his two screaming Mercury engines—we had a good look over a steep drop into the boiling cauldron of death. Powerful. Combined with the visual insanity, the juxtaposition of mechanical and natural sounds at extremely high levels made the experience almost overwhelming. "Pray to the God of Mercury!" yelled Mike through his larger-than-life grin.

For overall appreciation of a subject as large as Stikine, there is probably no better perspective than the airborne one. Although airplanes remove us from sound and other sensory elements of any given place, their speed and altitude capabilities help bring different locations together into one whole unit. How else is it possible to see meltwater sparkling from a headwater glacier and foaming through the Grand Canyon before shimmering into the ocean 640 kilometres away, all in the light of one day?

Putting a camera onto the airborne platform heightened awareness in ways often unexpected. Regardless of results, the camera lens demands focus on a given subject for extended periods of time, and sometimes a new perspective emerges. The technical results are not always encouraging: the colour can differ between identical film types from the same developer; the same stretch of river can look drastically different at different water levels; shooting at slightly different heights above ground makes for a consistently large problem; wheels and wingtips always want to be in the picture. These are but minor concerns, until attempting to put a series of images together in film-style format.

Nevertheless, the exercise was well worth the effort—days of S-turning back and forth with two 35-mm cameras clicking out a side window. Through repeated looks, we recognized the symmetry—headwater rivulets melting from a glacier juxtaposed with fibrous tidewater channels disappearing into the ocean—held together by a mighty strand of liquid motion. Here on the Rim of Fire, coastal granites and interior sediments are welded together by a volcanic band of igneous rock through which the wild, strong river slices a narrow path. How could any camera resist?

Site Z, the uppermost dam site of two in the mainstem canyon, is a dam-maker's dream come true: a narrow race of white water plunging between vertical walls about three hundred metres high. Wonderful to look at from a distance, it was probably less so for those working on-site: a nonstop roar in the ears and slight vibration underfoot in a place only briefly visited by the sun. While providing the only connection to highway base camps, helicopters were also the most practical form of vertical transportation between work sites at river level and trailer-town accommodations high above on the solid-rock plateau.

Tanzilla Gap, near the lower dam site, is exactly that: a very narrow gap between canyon walls where the Tanzilla River tributary enters over a six-metre waterfall. At least that's how it looks in low water … such as in 1985 when the first-ever white-water rafters in the Grand Canyon had to tilt their two-and-a-half-metre-wide craft on edge to slip through. At high water, it's a different story. No waterfall. No narrow gap. Only megatons of churning brown water in a typical section of the Grand Canyon. The dynamics are impressive at any time. Enhanced by the large catchment area upstream, the canyon water level is known to rise six metres in twenty-four hours; and, as the locals say, trees go in and kindling comes out.

Although canyon walls diminish in height below Tanzilla, the narrow width of the river channel varies little. The smaller side canyons are themselves spectacular while adding to the Grand Canyon's vertical grandeur. Immediately downriver from Tahltan Flat, Telegraph Road cuts a skinny ledge into the vertical wall on river-right and drops into a couple of creek crossings before levelling out on a bit of a plateau where we find YTX Telegraph Creek Airport (somewhat international) with its 5,000-foot (1,525-metre) runway of good-looking gravel … of little interest this day. Sawmill Lake on the other side of town was the recognized waterdrome for float-equipped aircraft such as ours: a small air-charter operator there was known to have a dock and a gas pump. Although known to host other types (Beavers and C185s), our Captain Bill, with a studied look and great wisdom, decided that at about 305 metres above sea level, the lake was too small for safe operation of his C180. We'd get in okay, but … whereupon the ever-attentive pilot not flying (a condition mandated by flight simulator instructors) made verbal reference to a photograph he had once seen showing two older float planes moored at the riverside landing in downtown Telegraph Creek.

With designs on a piece of Diane's renowned blueberry pie, the ever-competent pilot flying took us down for a look and a successful landing. However (big pause), conditions on the river immediately proved uncomfortable: a strong wind and an extremely strong current in a very confined space presaged the possibility of poor happenings, whereupon Captain Bill quickly kicked us around and deftly got us out of town. A demonstration of great airmanship. Once again, skill and cunning had won out over ignorance and superstition. We had once again gone where few had gone before.

Thank you, Bill.

Chattering and Clattering

6
Splitting and Splicing

In 1984, my meditation partner and new gallery mate, Mary (aka Prem Gatha), joined me for a pilgrimage to the southern hemisphere. We were off to see the wizard of Aussie who had been instrumental in the campaign to recently see the threat of hydroelectric dams removed from the Franklin River in Tasmania. Though details have long ago vanished, the aim of the journey was to connect with the river and some of the people involved in hopes of learning something. It was a whirlwind tour. After flying from Vancouver through Honolulu to Sydney, we travelled by bus to Canberra for a meeting with *whazizname* who was then well known for efforts on the anti-damming campaign. "Don't give up!" was the essence of his advice.

A bus to Melbourne for a regional turboprop got us to Launceston on the island of Tasmania for another bus to the Macquarie Harbour village of Strahan on the west coast, a place where tourism was beginning to supplant fisheries as the major economic engine. A small launch with an informative guide took our small group of tourists across the bay and then upriver for intimate looks at the lower Gordon and Franklin rivers, enough for us to appreciate their natural beauty and to celebrate their liberation from the threat of more dams—two were already in place on the upper Gordon. We came, we saw, and we touched the river. But did we learn anything? For me, the sight of numerous hydroelectric penstocks draped like spider legs over the island's central plateau was disturbingly motivational. For inspiration in the positive sense, it was only necessary to cruise the quotations in our souvenir guidebook:

> A river is the cosiest of friends. You must love it and live with it before you can know it.
> — *G. W. Curtis*
>
> Something will have gone out of us as a people if we let the remaining wilderness be destroyed; if we permit the last virgin forests to be turned into comic books and plastic cigarette cases; if we drive the few remaining members of the wild species into zoos or to extinction; if we pollute the last clean air and dirty the last clean streams and push our paved roads through the last of the silence … The reassurance that it is still there is good for our spiritual health even if we never once in ten years set foot in it.
> —*Wallace Stegner*
>
> God, the banker, does not deal in gold. His equity is life's respect for life … man must pursue his knowledge and understanding of nature not with the object of domination and the extension of power, but in a way in which awe and humility before the mystery of life and nature grow in equal proportion to his knowledge.
> —*Yehudi Menuhin, Patron of the Tasmanian Wilderness Society*

Back Home

From Stikine Country came reports of a substantial increase in activity around the Johnny Mountain gold mining prospect as well as a significant spike in interest surrounding Gulf Canada's 50,000-hectare coal property near Mount Klappan in the upper watershed. Gulf's 1984 prospectus detailed an open pit mine producing 3.5 million tons of anthracite per year while employing 1,100 people. Its townsite would require more direct road access and a permanent supply of electricity. Nearing completion of its Stage I submission, Gulf Canada had built a bridge across Little Klappan River and continued to upgrade the rail grade for access. In fact, a bulk sample of sixty thousand tons had been shipped out by truck for analysis and market testing. Stay tuned.

On the fifth anniversary of the first commercial fishery on the Stikine, and in harmony with the sockeye spawning cycle, the DFO disallowed salmon fishing on the river in 1984. Instead, comprehensive studies were undertaken to better determine extent of the fish stocks and their spawning sites. Under a grant of $180,000 to the non-profit Northwest Enhancement Society (Bob Gould et al. at Great Glacier Fishing) twenty fishers were hired to conduct the studies along both lower river mainstems and their tributaries as far up as Tanzilla. Though efforts to create a bilateral treaty had failed so far, the US authorities imposed a similar ban in its waters while anticipating a sharp decline in stocks. In parallel with Canadian studies upriver of the border, the Alaska Department of Fish and Game operated two sonar-equipped fish counting sites on their side … so went the quest to determine how many salmon go where. In the summer of '84, the Canadian DFO had asked for a one-year delay in the BC Ministry of Forests' decision on its impending land-use plan for the lower Stikine in order to complete all the fisheries studies. No luck. In October of '84, the BC Ministry of Forests approved additional logging along the lower Stikine River, and while claiming to incorporate the widespread objections heard from area residents, the Cassiar District's Ian Bowie stated, "I feel the forest resource can be utilized without detrimental effects to the fisheries resource and the recreation resource."

Local reaction was immediate and forthright. "We'll use any legal, political, and technical means available to us," declared Lynne Thunderstorm, RFFS coordinator, in announcing the group's plan to block implementation of the proposed logging. Referring to a recent meeting of area residents who strongly opposed the land-use plan, Thunderstorm suggested the ministry was using pseudo-public input as a means of rubber-stamping approval for whatever it wanted, despite some of its studies remaining incomplete. Residents for a Free-Flowing Stikine thereby withdrew further participation in the public involvement process of the BC Ministry of Forests. Asserting his personal environmental sensitivity, Ray Halvorson of HAL–PAC suggested the RFFS's concern was beyond environmental: "The want to reserve the Stikine River valley for a very small group of people, for their own personal enjoyment." Soon thereafter, HAL–PAC was awarded harvesting rights to seventy thousand cubic metres of cottonwood and spruce on approximately 350 hectares at the confluence of the Stikine and Katete rivers, slightly downstream of its previous cut-block on the Iskut.

About this time, Friends of the Stikine in Vancouver welcomed a new member: Grant Copeland.[9] An environmental planner with successful conservation campaigns elsewhere in the province, he wanted to contribute his energy and experience toward suitable protection for the Stikine. We had heard of him and of his colleague-in-arms Colleen McCrory through their involvement with recent creation of Valhalla Provincial Park and with ongoing protective efforts for Gwaii Haanas (South Moresby Island), part of Haida Gwaii (formerly Queen Charlotte Islands). Grant had certainly heard of us, and he had already tasted the Stikine: he was in that previously reported photo of Jim Bourquin's raft at Fountain Rapids in '79. We began working together immediately. The masthead of 1985 newsletters (#11 and #12) show Friends of the Stikine to be then located at the home of May and Jim Murray—1405 Doran Road, North Vancouver BC, V7K 1N1—with Nancy LeBlond as president and editor, Peter Rowlands as vice president, and May Murray as treasurer and membership secretary. Listed directors were Sylvia Albright, Kriss Boggild, Gary Fiegehen, Rosemary Fox, Carol Lambert, Don McClure, and Phil Lander. Rosemary and Irving Fox had moved to Smithers.

The Convention

Although our modes of travel remain a mystery, we were there: Grant Copeland, Gary Fiegehen, and me in Telegraph Creek representing Friends of the Stikine for a two-day workshop on May 22–23, 1985. Although names of the instigators and the organizers of this event have been lost to history, it's probably a safe bet that Grant Copeland of FOS and Lynne Thunderstorm of RFFS were in on the ground floor of this interdisciplinary exploration of land-use management options in the Stikine watershed. For the first time, several conservation groups with common purpose were able to get together and to meet with representatives of two levels of government and with the local people most concerned. I unofficially dubbed it "The Telegraph Creek Convention."

The Association of United Tahltans as de facto hosts were present in goodly number to hear from Parks Canada (Ottawa) and DFO (Whitehorse), the BC Ministry of Lands, Parks, and Housing, as well as the BC Ministry of Forests. RFFS and FOS were joined by Yukon Conservation Society (YCS) and Southeast Alaska Conservation Council (SEACC). A few impressions remain from that community-centre gathering in the main village. Hearing Grant's introductory thanks for being welcomed into Tahltan territory was a personal first for me ... social progress from the days of Robert Campbell when HBC traders claimed propriety over everything in sight. Around the room, hard-nosed aggressive/defensive posturing by the United Tahltans was also palpable: they were probably already tired of hearing airheaded strangers telling them how to manage their lands. Miners and loggers gave you work; they didn't put new colours on your map and strange names on your territory. The most prominent spokesperson among the Tahltan delegation was a lawyer named George Asp.

For us tourists, the lodging at RiverSong provided for more convivial conversation on a greater variety of topics beyond the common thread of concern weaving through our conservation-minded acronyms. In an undefined and unspoken manner, we recognized ourselves as a team without need of a leader or any holy grail. Though never discussed as such, underlying our lighthearted banter was hope that one river left to run free in one small corner of our planet might open a door to another way of looking at our world. Meantime, it was a pleasure being in the company of new friends Mary Ellen Cuthbertson, John Sisk, and Bart Koehler of SEACC, joined by Tom Munson of YCS. A warm-hearted riverboat ride to Wrangell with Dan and Dave brought fitting closure, getting many of us together on the same river—and on the same page.

A subsequent press release, with appropriate quotes by Lynne, Grant, and Mary Ellen outlined the resolutions agreed upon by the conservation groups in attendance.

> 1. To protect and maintain the integrity of the entire Stikine River watershed.
> 2. Support for the settlement of the Tahltan aboriginal land claims in the watershed.
> 3. Recommendation for National Park designation for major portions of the Stikine River subject to settlement of Tahltan land claims.

Along with upper-river coal mining, mid-river damming, and lower-river logging, the quality-of-life issue for watershed inhabitants completed our list of major concerns. Calling for immediate moratorium on further alienation of agricultural and residential lands as well as on access roads and flooding reserves, our Telegraph Creek Convention of 1985 also recommended against any commercial logging on the lower river and against all hydroelectric development as well as against any large-scale mining initiative—all pending resolution of the Tahltan land question. Similarly, any and all land-use proposals and designations should be subject to the Tahltan land claim.

It was also suggested the international, worldwide significance of the Stikine watershed might be best recognized by its nomination for Biosphere Reserve status according to guidelines of the International Union for Conservation of Native and Natural Resources (IUCN).

Park-like protection of some form was recommended for the entire main river corridor. Though the idea of a provincial park was not entirely dismissed, the extremely low probability of its implementation helped encourage overall preference for something in the *national* park tradition. Although a six-hundred-kilometre-long park didn't make sense on anyone's map, our call for a national park *reserve* would put a freeze on all industrial development along the river while allowing time and space for resolution of the Tahltan land question. If then so desired, a national park could be created around the Grand Canyon of the Stikine on a scale suitable to and employing Tahltan people at all levels. In any event, all traditional activities, including trophy hunting, would be allowed to continue. Even if a national park did not evolve, national park reserve status would protect the river's natural state while maintaining its eligibility for heritage river designation and/or for any other program that might appear.

My personal favourite recommendation was for creation of a *headwaters* national park reserve (separate from the river reserve) encompassing the headwaters of the Spatsizi, Klappan, Nass, and Skeena rivers to join with Stikine in a set of high-elevation source waters exempt from industrial contamination. By filling in the space between Spatsizi and Tatlatui provincial parks, such a reserve could contribute to a large contiguous area for wildlife while protecting known caribou calving areas. All boundaries would be subject to resolution of all neighbouring Indigenous land issues, and all traditional activities, including guide outfitting, would be allowed to continue.

Solo

Personal challenge, along with the need for storyboard photography, inspired my solo adventure in that summer of 1985. (The idea of a motion picture to stimulate Stikine awareness was still going strong.) Having room to spare in my chartered Beaver, RFFS coordinator Lynne Thunderstorm and her neighbour Jackie Williams (Willie's brother) accepted an invitation for a chance to see something of the Spatsizi Plateau and the upper river. At my request, we offloaded at a clearing on the southeast corner of Tuaton Lake to enjoy a stretch of shore time and a couple of stories from Jackie, one of which suggested the name "Tuaton" referred to the human hand because of its five principal feeder streams.

From the shoreline, a gentle rise of yellow and purple wildflowers and a lone spruce offered an ideal camp site facing west across the lake and up the main Stikine tributary. Through the scrub birch and willow immediately behind was my target—GP Creek, so named by Gary Fiegehen during our photographic mission to its headwaters three years prior, GP being his convenient amalgamation of my birth name and my spiritual name (Guruprem). It hadn't taken long thereafter, with a little help from our friends, for these initials to expand in possibility: through Gruesome Pilot, Paddler, Poet, and Photographer to Grizzly Pete on his way to becoming Grampa Pete. This creek of interest is the same glacier-fed stream I first visited in 1979 with Hal Marsden, the one flowing into the Tuaton Lake narrows from the snowfield col about seven kilometres south, on the west shoulder of the unnamed peak 7429—aka "Mt. Marsden" at the head of "GP Creek."

Whatever its proper name might be in the tongues of Indigenous peoples or in the later lingo of guide outfitters, GP Creek is a short and easily accessible headwater tributary for capturing on photographic film the lively complexities of a newborn river, the sort of detail unavailable to my envisioned airborne platform (drones being then unknown). Off I went: my two too short legs tangling with the buck brush while carrying two Canon 35-mm SLR camera bodies and a couple of lenses, a small overnight backpack with a bit of grub, a sleeping bag, a rain poncho, and a .308 calibre rifle, just in case.

No, although black bears were often seen on riverbanks, I never caught sight of the feared grizzly bear on this trip or on any one of numerous extended treks in the upper-river high-country over a ten-year period. Lucky perhaps; maybe just noisy and smelly. Given the amount of extended care it demanded and the amount of extra weight it added, I only ever carried the firearm on this one Stikine trip. On my way up the creek, caribou were often in the picture, sometimes curious but always keeping their distance. The weather held. At the base of a large, warm boulder on the second bench from the top, sleep came easily … with the rifle by my side.

Dawn was a dream. No wind. No cloud. The sound of water giggling over pebbles and gurgling over rocks. Traces of petalled colour in the softened grasses of an alpine meadow looking down the length of valley-bounded stream disappearing into a silver-slivered lake before a horizon of purple mountain tops. Behind me: a broad arc of steely black peaks backdropping a tall, skinny waterfall bisecting an attractively patterned wall of dark earth and melting ice. Peace. Surrounded by nature's beauty and enjoying a state of serenity, the awareness of other beings came slowly and gently. A dozen or so caribou of various size were calmly and contentedly grazing everywhere about me, some within fifty feet. No fear anywhere. Only awe. CLICK. The sound of my camera shutter alerted my new friends to a mechanical presence and off they flew, later to be seen down valley, again maintaining proper social distancing.

The downstream hike was an all-day affair, taking my matchstick-driven curb-gutter river-runs from childhood to a different level: inching across narrow snow bridges and white-knuckling down sharp little canyons, standing atop slippery rocks, and kneeling in squishy greenery, all the while remembering Gary's first dictum of professional photography when seeking the ideal shot—you never get the best one. It was worth it. I collected a sufficient number of satisfactory images for storyboard purposes along with greater understanding of liquid dynamics and its geological influences. Again, the act of looking through the lens gave more than the picture taken. No longer just groundwork for a personal attempt to counter politico-industrialism, this time "up the creek" became a meditation in motion—slowly ever forward, turning and dropping in harmony with a perpetual flow, liquid crescendos floating on a greater silence. Site Zed and its brethren dam sites were no longer factors in my life. This moment with the young river was all that mattered.

Skeena ... Omineca ... Cassiar ... Stikine
Mountain families crowned in bright crystal
Great robes of ermine snow reflecting light
Towering spires adorned in glaciered ice
Volcanic cones awash in silvered sheen
Melting ever downward in liquid perfection

Dripping ... dropping ... falling ... flowing
Seeping and sliding through earth and stone
Dribbling and drizzling into rills and rivulets
Trickling and twisting into puddles and ponds
Spilling and splashing into brooks and streams
Carving and cascading into lakes and tarns

Shimmering ... shining ... glimmering ... gleaming
Sparkling diamonds in alpine splendour
Resting and reflecting, rippling and rejoicing
Pushing and pulsing, streaming and squeezing
Through the gnarly narrows of Fountain Rapids
Plunging and pouring to a river running free

Travelling alone has its advantages, most notably the absence of any need to seek agreement on timing or direction: all possibilities were possible, and there was never immediate need for a decision; simply sit down and think it through. The downside, of course, is the absence of a responsible adult to verify unusual perceptions and to safety-net unusual acts of bravery attempted in a state of ever-common absentmindedness. It was fun. My love of life soon inspired greater awareness and meticulous care in the use of knife, axe, and rifle … and there was always someone to talk with.

The type of in-depth study conducted on the headwater creek carried over to Fountain Rapids where a full day was dedicated to the task. Aided by experience of two prior occasions, I knew the low-lying spruce grove at head of the portage would be a relatively damp place to camp, obviously affected by waterfall mist and probably by a back-eddy of wind in the tight terrain. By contrast, my two-night camp at foot of the portage was on a high-and-dry site among a scattering of bright pines. Yes, indeed, the journey continues, and the education never ends. "Too soon old and too late smart" was another common quip from my father.

My one-day commitment to Fountain Rapids included a short canoe crossing above the entrance for my first-ever exploration of the far side river-left clifftop before returning for the usual waterline inspection on the near shore. I took scores of photographs from every attainable perspective. The water seemed higher: the rock shelf last seen sporting an abandoned, high-and-dry life jacket was now completely under water. Big water. In fact, smack dab in the middle of the churning mass was an especially tall boil of white water burbling high above the rest. Looking for all the world like an old-school drinking fountain, it may be responsible for the rapid's name. Resuming my river journey below Fountain, I bounced lightly downriver, joyfully playing with scattered blobs of white while capturing storyboard images. WHAM!

Chapea, a barely-before-noticed creek came in with such fury that its previously benign rock garden had become a frothing mass of crosscurrents among a strange-looking set of boulders. Surprise! My lingering question about water level was answered in a flash. Forced to an outside corner by lack of technique, my novice-level paddling status was confirmed by eager hands clutching willow branches. When in doubt, sit down, take a rest, and think it through. The playful young river of '79 was now a surging torrent filled with life-affirming moments of near death, substantially higher than it was with the boys in '82. Obviously, lake-level differentials are difficult to determine and probably not always indicative of what's happening down below. Fortunately, on this occasion, a shoreline scouting mission helped determine a paddling route down to the portage trail around Chapea Rapids proper.

Farther along at Metsantan, while manoeuvring onshore to better photograph my pet-rock talisman, the sight of debarked pine trees stopped me in my tracks: a metre-high vertical strip of bark was missing from the upstream side of virtually every tree trunk along that entire shoreline bench, itself about a metre above the river surface. Maybe ice? A post-trip discussion with local folks confirmed my observations to be the normal side effect from dramatic break-up of heavy-season ice … and, as later informed, that season's water level was well above average.

Oh, good to know. Thanks. Along with my attempt to visualize the magnitude of such an ice-filled event at "Calamity Corner" came an ever-humbling increase in appreciation for watershed dynamics. Yes, it is impossible to step into the same river twice. And yes, the path to wisdom begins with knowing you know nothing.

The next planned stop was Adoogacho Creek, halfway between Metsantan and Chukachida on river-right, across from the plateau, where a tall, white slash in a sea of green had caught my eye during aerial photo missions of the previous two years. No longer just another line on the map, Adoogacho Creek had become a point of interest by way of a sizeable waterfall close to the Stikine but out of sight from the river. A scouting mission there took me directly onto a height of land for a fine sight: a moderate stream twisting through a rock-bound crevasse and plunging about twenty meters into a dark pit of unforgiving boulders and billowing mist. The stream's raging torrent regained composure in the lowland forest before meeting the mainstem river. Above the thundering cascade, the incoming stream emerged from a forest of moderately spaced trees upsloping with moderate pitch toward the alpine. Hmm.

Back on the river with *Dimples*, a customary gaze up the valley of the Chukachida saw me around the northeast corner and into camp at Sanabar Creek before rain arrived but arrive it did. Tent-bound with ample time to think about such things, my focus rested on this particular place and more particularly on its name. It was a great campsite, one used on each of three upper-river trips to date: easy access through a gravel-bar eddy below a lively stream, room for several tents on a level bench of vegetated sandy soil looking northeast across the creek mouth, good drinking water, trees nearby for cook tarps and wind protection, and a forest of standing deadwood available for drying boots and clothing around open fires. A deserted trapper's cabin slightly upstream on the creek gave silent approval of the location.

Hmm, s-a-n-a-b-a-r … if the middle *a* was changed to a *d,* we would have *Sandbar* Creek, a truly fitting name for the place in keeping with its appearance. Who knows, maybe a smudged pencil mark in a surveyor's log or a typographical error in translation was responsible for creating a new word. Then again, maybe Sanabar is an Indigenous term known only to a few. *Enough. Get some sleep.* Orographic lift from the bordering mountain range is probably responsible for the localized rainfall here at the northeast corner of the plateau. The river will probably be even higher tomorrow.

The next morning, it hit me with a wallop. Jewel Canyon! Riding a strong current and getting close, eyes and ears were scanning for any sign of excitement while my mind queried the possibility of a portage trail. Suddenly, it was too late for the latter option if it had ever existed. Without any real warning, the fast-moving river had become a mass of roiling water with barely a rock in site, just an endless and disorganized array of wave tops leaving no option but to go with the flow while avoiding the biggest splashes of white. Fighting the current toward either shore seemed like a losing proposition—memories of being caught in high winds on a finger-narrow lake with no way out. Never simple, the challenge of manoeuvring through rocky obstacles and strong currents is significantly more trying when conducted solely from the far-aft stern seat of any canoe. The normal practice of changing seats (positions) and paddling "backwards" from the front seat was made near-impossible in my old Grumman by a tubular thwart located immediately behind and slightly higher than the bow seat. As such, the lazy-comfort technique—paddling from the stern seat with the load biased forward for balance—had been thus far satisfactory.

Of no great surprise, retrospect suggests the advisability of repositioning the load and paddling from the knees while positioned in front of the awkward thwart. This written sentence is living proof that good fortune had joined up with naïve determination to invoke a successful outcome. Without question, memories of my companions' "swim" here in '82 strongly influenced caution to whatever extent possible. Without back-up assistance, Gary and Monty would have had a tough time surviving the effects of hypothermia in order to regain their downstream track. Although they retained possession of their boat and food, they were one hundred kilometres from everywhere on an unknown river without a map or any form of outside communication. Excitement and success are but small players in a greater game. Survival odds for one person and an upset canoe in a situation such as theirs are not encouraging. It is often better to be lucky than good.

Although, in the moment, luck didn't seem to be playing a major role here in 1985, it was still present, and Jewel Canyon did trigger greater regard for that old saying while prompting a closer look at the relationship between good fortune and personal ability. While great skill is known to create its own opportunities for lucky breaks, my experience here differs—good fortune allowed the river to abide mediocre ability while promoting greater understanding. I was lucky.

Any hope of an early camp at the Pitman confluence, thirty kilometres downriver, was again wiped out by a look at the place. As it was in '82, my fondly remembered site from '79 was long gone. Obliterated. All that remained was increased awe for river dynamics. A first-time viewer would have no sense of what was once there. Now, in behind this "Photoshopped" scene was another reminder of nature's overwhelming speed and power—a vast "new" forest of fire-blackened spires.

YIKES! Beggerlay Canyon came up really big and really fast—far grander and certainly more threatening than it was for the four river rats in '82. With the high-water state of the mainstem, and regardless of the creek's flow, most of the undesirable canyon-mouth territory was drowned out by big waves and crosscurrents all squeezing in with great vigour, making the area even more undesirable. Although the river-left chute seemed to be wider, it was also stronger, and the high-water rollers in the canyon looked HUGE. Out front, the noise level was high, and time was short.

As if there was much choice, it took a degree of determined faith to aim for the boulders just upstream of the raging torrent coming in from the right. Yes, the eddy! Home safe and sound. The boulders were still big, the creek was still loud, and the canyon was even more intimidating. The place is amazing at any water level. As the last planned camp on the upper river, Beggerlay invited a stay of two nights in order to assimilate the overall trip experience while getting psyched up in preparation for re-entry into civilization. It was also an excellent opportunity for admiring the river's dynamic, for studying its movements and for listening to its rhythm … feeling its vibe and trying to be one with it … studying the route to be taken and working up the courage to try it.

With thanks again to Bill Mason's teachings, my previously tested paddle-turn-paddle-turn-brace procedure again proved successful for entering the canyon. Inside, however, the strength of whirling pool eddies and the height of rolling round waves were enough to test the mettle of any solo paddler in an old tin boat, no matter which seat he was paddling from. After another significant blip in my heart rate at Little Beggerlay came my first contact with human beings in two and a half weeks: two single-seat airboats (swamp buggies) powering upstream with aircraft engines and pusher props. Not impressed. After trucking around to meet folks in Telegraph Creek, I returned to reading my travel book of choice, which suggested that by being three weeks alone, it's sometimes possible for a person to attain another level of consciousness. Hmm, perhaps, a few more days at Beggerlay might have done the trick ….

On the drive out, I met a large windfall aspen tree spanning the width of Telegraph Road. Hmm, nuthin' to it. Using its heavy-duty front bumper with the winch on top, Brutus will just push it off to one side … WHAP! The main stem of the tree, which hadn't broken off slammed across my windshield like a sprung rat trap. Fortunately, other than character cracks across the front window, there was no damage to the vehicle—and no discolouration to any item of my attire. So much for that next level of consciousness. Maybe that's why the locals carry chain saws as well as two spare tires in their trucks. For the record, Brutus my trusty Landcruiser was now three for three: three trips on the beautiful Telegraph Road and three short stories of the mechanical genre.

7
Singing and Soaring

Having survived my solo test flight, it was time to join a team of like-minded aspirants heading for serious-minded discussions. The 1985 Heritage for Tomorrow Canadian Assembly on National Parks and Protected Areas took place in Banff from September 4th to 8th as the culmination of a comprehensive two-year public input planning process conducted Canada-wide. A strong British Columbia caucus led by Colleen McCrory, Vicky Husband, and Grant Copeland included Lynne Thunderstorm from RFFS as well as Irving Fox and me from FOS to provide additional voices for the Stikine. A special ad hoc workshop was formed to work on urgent proposals for South Moresby Island and for the Stikine River, which were subsequently presented to the final plenary session by Colleen and Grant. A standing ovation response by the entire assembly was highly gratifying; included in the audience were new Minister of Environment Tom McMillan and several senior officials from Parks Canada as well as several provincial park ministers.

Along with watershed values put forth by our Telegraph Creek Convention, this forum identified the upper Spatsizi–Stikine plateau to be in the so far unrepresented Region #7 of the Canadian national parks system. More importantly, while calling for a national park reserve on the entire Canadian portion of the river, this *Banff Manifesto* stressed that any such designation must be subject to resolution of the land claim of the Tahltan Tribal Association, that their traditional activities of hunting, fishing, and trapping be allowed to continue, and that the Tahltan people be involved in the planning and management of any proposed park. Though never enthralled by the idea of a river-length park, I happily went along with whatever the proven professionals recommended. Not necessarily a park, but a park if necessary.

Groups from across BC joined in support of recommendations for the following: a South Moresby National Park Reserve and a South Moresby National Marine Park Reserve; a Stikine National Park Reserve; wilderness protection of the Stein River Valley; Tribal Park designation for Meares Island; wilderness protection of the Khutzeymateen Valley; a Cascade wilderness extension to Manning Provincial Park. Our summary pointed out that fewer than one-half of the fifty-two natural regions in BC were represented in the provincial or national parks systems, and if all the park proposals in BC were accepted, more than ninety per cent of provincial lands would remain available for resource extraction. While the South Moresby and Stikine proposals were being recognized for their national and international significance, the absence of BC Parks Minister Tony Brummet from the assembly was not encouraging.

Back to Business

As noted in the FOS newsletter of November 1985, a letter from the Cassiar Forest District clearly stated the ministry's intention to continue logging along the lower Stikine despite overwhelming opposition by local people at its own public meeting, and despite our Telegraph Creek Convention's recent call for a moratorium. It seems the $16,000 stumpage fee—for whole logs exported overseas at forty cents per cubic metre without any Canadian processing—was being offset by a $65,000 non-manufacturing levy collected by MOF. Readers dismayed to see this occurring in the region John Muir described as "Yosemite 100 miles long" were being encouraged to write to BC Environment Minister Pelton and BC Forest Minister Waterland as well as Environment Minister McMillan in Ottawa.

In a letter dated January 16, 1986, to BC's relatively new Wilderness Advisory Committee (WAC), RFFS Coordinator Lynne Thunderstorm addressed the contradictory nature of the provincial government's attempt to deal with rising wilderness preservation concerns among the general populace. Sharing a long list of past contributions to the public review process, in time and paperwork, Lynne strongly suggested it was wasted energy, nothing but token public input to justify a predetermined decision. How could a balanced management plan be arrived at in three short months by a government committee which included but one pro-environment person and not one Indigenous person? Is it reasonable to expect a citizens' group to apply their unpaid labour to a paid panel that will ultimately ignore it? "At the very least, a logging moratorium should be put in place immediately, until final and firm decisions, subject to public approval, are made," she said, with the lower Stikine logging in mind, "… we will not legitimize a process that is clearly not for the public good by making a submission."

The recommendations of the Wilderness Advisory Committee which were released after three months of public input were heartily denounced by conservation groups province wide. In almost every case, logging, mining, and hydroelectric development were allowed to continue with no particular attention given to wilderness protection or wilderness values. At the time, "wilderness" was simply the buzzword for land. While stating "it sought recommendations that implied no net loss to dislocated resource users," the committee was *not* permitted to consider the net loss to the tourism industry and was instructed *not* to address outstanding Indigenous land claims. None of the WAC members visited the Stikine, and their report on the area failed to reference its national and international significance as recently highlighted. While failing to reference the Mt. Klappan coal project with its washing unit already in operation on the unprotected headwaters of Spatsizi River, the report goes on to recommend a multiple-use recreation corridor management plan for the Grand Canyon of the Stikine whereby dams and hydroelectric development could continue, and for the lower river a scenic corridor containing environmentally sensitive clear cuts (i.e., out of sight from the river).

Seeking Public Support

Yes, even for those immersed in its issues, there was a life outside of Stikine. After several years of flirtation with large airplanes and exotic foreign destinations, common sense and diminishing need lured me home to the much-loved B737–200, Boeing's two-engine airliner, the small "sportscar" that had first taken me around BC District (a collection of short runways in the middle of nowhere) and had introduced my all-time favourite route to fly: Vancouver to Whitehorse. There were other interesting ways of getting north to Yukon, but none were as scenically outstanding (or potentially turbulent) as the one along the coast range. It was wonderful to be back. Unbeknownst at the time, this peppy little two-engine runabout would become my lifetime favourite aircraft and would be happily flown until it was retired from service twelve years later. About the time of my upgrade to left-seat captaincy on the airplane came my election to the presidency (later chairperson) of our small conservation group, Friends of the Stikine. Both promotions brought increased responsibility offset by more freedom of action.

One fine day in Vancouver, photographer-friend Gary introduced me to a gentleman by the name of Blair Shakell, the owner–operator of Scribbler's Inc., a frontline audio-visual production house, who wanted to donate his expertise to our cause—a narrated multi-projector slide show on videotape was his idea. Sensing their enthusiasm and seeing how far these two were already into it, there was really no decision to be made: it was already happening. "Go for it!" blurted my mouth. "Our friends at the next meeting will think it's a great idea … with any luck." As it turned out, it *was* a great idea. Although it came close to blowing us away with more volume and drama than anticipated, we weren't about to argue with a proven master of the medium.

A Call to Action: Save the Stikine was exactly that—a fast-paced description of a dire situation with explicit instructions for doing what needed to be done while identifying the government agencies in need of encouragement. Set to lively music and artfully narrated by a retired radio personality, this nine-minute "heads-up" featured about two hundred of Gary's Stikine slides, mixed and matched in rapid-fire succession from six highly synchronized slide projectors. It was wonderful. Captured on VHS videotape for ease of transport and presentation, it was even more wonderful. It was so wonderful it won a 1987 Gold Quill award from the International Association of Business Communicators

(IABC) from among 2,914 entries. Created for the purpose of increasing public awareness, *A Call to Action: Save the Stikine* was a mini masterpiece from a maestro with a big heart. Thank you, Blair.

By then, FOS was well committed to the pursuit of public support. Gary's iconic Sergief Island photograph had become a dynamite poster, serving well as a promo-piece business card as well as a fundraiser. A series of postcards featuring Grand Canyon goats and other Stikine images were also available. If there was a fault in our gung-ho approach, it was over-production—too many items for a small market, though beating the bushes big time with impressive visuals and formidable rhetoric. We (mostly me) may have become a nuisance at times. The well-intentioned folks at the BC Outdoor Recreation Council (ORC), under the illusion of there being other areas of concern in the province, must have been shaking their heads in bewilderment: they didn't realize we were with Jake and Elwood, and we were on a mission from God. It was probably Nora Layard, in charge of patient care at ORC, who gave me the heads-up about the conference at UBC, which her organization was putting together for the spring of '86. Is this just another token discussion about the issues before getting back to business as usual? No, she said, it's much more than that: it's an international trail and river conference with speakers from around the world. An expensive talkathon? No, not too pricey … with lots of mini groups and workshops. Can we get Stikine on the agenda? No, but there will be a speaker's corner where you can say your piece and a gift store for selling your wares … and you can lobby your brains out if you want. Here's the program announcement:

1986
INTERNATIONAL CONGRESS ON TRAIL AND RIVER RECREATION
May 31–June 4
Vancouver, British Columbia, Canada

Incorporating the 8th National Trails Symposium
of the
National Trails Council, United States of America

Congress Developed and Coordinated by
Outdoor Recreation Council of British Columbia

Juri Peepre, chair of ORC and congress chair made the following introduction:

> *From a Canadian perspective, the conference is really about our heritage, for it was the rivers, portages, and trails which for centuries provided first native and then European trade routes … We now live in a more complex age, where motives for the exploration of rivers and trails are linked to an inner spiritual and physical need, and not to the service of a foreign king or commercial empire … We are here to talk about the conservation of our trail and river heritage, yet the focus of the conference is on people … Trail and river recreation is fundamental to our way*

of life and we must present our best ideas to governments and other decision makers to improve the protection and management of the land which provides these opportunities ... Trail and river travel in Canada has, and always will be, a part of our way of life. The great challenge to our society, and others around the globe, is to preserve a part of these experiences much as the original inhabitants and adventurers found them.

With a dozen well-lettered speakers and hundreds of delegates from around the world, this first-ever international trail and river congress was a winner from the get-go. Obviously, the province's Outdoor Recreation Committee had done a magnificent job putting it all together and, obviously, Stikine had to be there. On my wing were Grant Copeland (of recent Haida Gwaii and Valhalla Park success) and Dan Pakula (river guide and lodge operator in Telegraph Creek).

For opening night on the Saturday, FOS rented the 6,400-square-foot ballroom in the Student Union Building for a Stikine Celebration dance featuring two bands—the Ogedengbe Drummers (Dido and other hot hands) and Saint Nodisco (Ian McConkey and friends)—$5 at the door and prizes galore (kids free). For the lucky winners: two days for two people at the RiverSong in Telegraph Creek with a two-hour boat cruise of the lower Grand Canyon courtesy of Dan Pakula and friends; one hand-knit hooded sweater donated by someone; Great River STIKINE posters, framed and unframed; and postcards by the handful. The music was great, of course, and everyone who attended had a good time. There was certainly plenty of room to dance. By the time conference delegates escaped the free booze at the official welcoming session across the street, we were having to shut down because of noise restrictions. My career as a dance promoter got off to a slow start. But we weren't done yet.

For the second day's post-conference evening activity, FOS hired an aircraft from Skycom to repeatedly overfly the harbour tour ship MV *Britannia* that would be carrying a full load of congress delegates on a sightseeing excursion under the Lions Gate Bridge and around Point Grey to the music of a Dixieland band while being nourished with food and drink. The chartered aircraft would be towing a banner that read, **BEGIN THE BEGUINE—SAVE THE STIKINE**, with the intention of creating interest in a subject outside of the conference's strict agenda. Brilliant. Under clear skies, our boatload of happy delegates departed Coal Harbour by the Westin Bayshore hotel and turned to starboard, instead of to port as was expected. The captain's voice came over the public address system advising: "Because of high winds and waves offshore, we would be motoring east and north up Indian Arm in the interests of passenger comfort."

Gasp. Only one passenger onboard was uncomfortable with that decision, and we knew who he was. A desperate visit to the ship's bridge found only a radio telephone for in-harbour communications. (What's a cellphone?) The banner-towing pilot was probably a youngish newbie building time in his/her logbook with hopes of something higher and faster. In any case, it seems the airplane driver didn't know what our vessel looked like and/or didn't know where to look, assuming he or she tried. Alas, along with some of the weekend sailors returning to port, the only international trail and river delegates who might have got the message were those taking in the sights at Wreck Beach. My career as an air traffic controller got off to a low start. But we weren't done yet.

As well as lobbying our brains out (a marginally productive activity for someone so short of tact and diplomacy) *The Stikine,* our mission from God, made it onto the printed activity calendar as an inclusion in both scheduled speakers' corners where, by default, Grant Copeland presided as our designated spokesperson. Although we could not coerce our audio-visual *A Call to Action* onto the program for a general viewing, it was offered as a side option during the regular afternoon break on Monday's day two in concert with an established media room function. Although few people were able to escape workshop activity in time to see it, many throughout the building heard its message loud and clear: "THIS IS A CALL TO ACTION!" Someone (not me) must have inadvertently mis-set the volume control for our already high-volume piece. Not my problem.

Tuesday's day three was entirely devoted to field trips in the region: Mount Baker, Skagit Valley, Whistler, North Shore hiking, Chilliwack Valley, and Chilliwack River rafting. As wonderful teasers for would-be attendees, some such events are inevitably cancelled because of weather or lack of a quorum; such was the case for my river-rafting preference. But we weren't done yet.

For several weeks prior, we Stikeeners had been kicking around the fantasy of a field trip to the river. Really? Well, if there was an airplane available at the right time, and if we paid for the fuel, and if the crew worked for free … and if the weather was good … it might just be possible. Preliminary discussion on the matter with my employer and one of the sponsors for the congress, CP Air (Canadian Pacific Airlines) had been fruitful: CPA had recently acquired EPA (Eastern Provincial Airways) and had added additional Boeing 737 aircraft to its fleet, along with an injection of down-home good humour from the Maritimes. In fact, Dave Wheeler, one of those newer managers in the flight operations department, shared our field-trip fantasy. As such, it was agreed that for the sum of $10,618.00 to cover fuel costs, FOS could charter a B737–200 aircraft during the early morning hours of Wednesday June 4, 1986. We already knew who one of the pilots would be.

Motivated by fear of my own in-flight fantasies—a high-speed low pass over the river in front of the RiverSong Café and General Store, followed by a steep climbing turn to set up a low-speed pass over the Telegraph Creek airstrip with everything hanging, full flaps and wheels down and locked, as an educational display for the school kids—the search began for a responsible co-pilot with a sense of adventure. From among numerous willing volunteers came Garry Grant, a check pilot, a supervisory figure from within the administrative sector of the respected Department of Flight Operations. Good choice. Garry was the kind of guy who would cover my back and take shit from no one while keeping me out of trouble with the boss. Customer service director Leanne Chambers (later Niewerth), known to be sympathetic to our cause, was asked to be responsible for passengers in the main cabin customer seating area, for which she agreed to assemble a pro bono back-end crew. The long-range weather forecast was encouraging, and we started selling tickets.

For the price of $125, conference delegates were invited to reserve seats on a possible flight to somewhere called Stikine, scheduled to leave YVR at 0300 hours on June 4th. Interestingly, the first person to sign up was keynote speaker Dr. Roderick Nash from the University of California. Close behind were primary speakers Bing Lucas, chair of the New Zealand Walkway Commission, and Dr. Paul De Knop from Free University of Brussels. Rounding out the top-of-list keeners were delegates from Seattle, Washington, Austin, Texas, and Tallahassee, Florida. Encouraging, but we had a long way to go: a full house of ninety-six souls on board would barely cover the cost of the gas, especially with free seats being allocated for immediate family and guest of honour Dan Pakula of Telegraph Creek. Free seats were also available for supernumerary crew members: official photographer Gary Fiegehen, assigned journalist Monty Bassett, and relief pilot Grant Copeland (a former USN pilot who would occupy the cockpit jump seat). After allowing as much time as possible for congress-goers to sign up, we began flogging seats in the outside world (where we found limited success) while also offering free seats to a handful of media personnel who were known to be aware of our environmental concerns.

Despite a determined sales pitch and intense lobbying efforts, we were desperately short of a full airplane on the day before the planned flight; it seems some people were convinced our field trip was only a public relations gambit to gain exposure for the Stikine issue without serious plans of actually getting airborne. On the afternoon of June 3rd, we were still about thirty seats (and a lot of gas money) short of a full load. "Don't do it, Peter; it's not worth it," came Grant's wise counsel at the eleventh hour.

Not quite ready to give up, a drive out to the company hangar at YVR gave me a last look at the weather. Amazing! The dreaded big L, the customary low-pressure system often hanging out in the Gulf of Alaska was nowhere to be seen. Yes, a big patch of grey wetness off to the southwest was expected to reach Vancouver early next day (tomorrow) before spreading northward; however, at the moment, there was not a cloud in the sky anywhere between Anchorage and Prince Rupert, and it was forecast to remain that way for the next twenty-four hours. There was no threat of weather reaching Stikine until long after our planned return. Severe clear and flat smooth … no bumpy winds. What's not to like? With barely a blip in my heart rate, my gut was telling me to go for it. *Too good an opportunity to pass up. Do it for you own selfish reasons. It's only money.*

"By the way," added the on-duty flight dispatcher, "there's been another equipment change and you've now got 720."

Recognizing one of my favourite tail numbers, a long-ranger with an extra fuel tank in its belly, my internal conversation resumed: *C'mon Peter, one of the finest-flying airplanes known to man, five thousand gallons of gas in the tanks, thirty thousand horsepower on the wings and not a cloud in the sky. You'd be insane not to go.* "Okay, fill 'er up!" announced my acceptance of the flight plan. "Thanks for your help … we're outa here!"

Flying Time

At 0200 hours on June 4, 1986, airport check-in was a breeze. Friendly agents. No hassle. No tickets. No security. No checked baggage. Only gaggles of chattering night owls casually strolling toward Gate 22, a meet-and-greet in motion. Along with about fifty unknown congress-goers, familiar faces appeared: Friends of the Stikine stalwarts May Murray and Bill Horswill; the long-pestered river lovers of ORC—Nora Layard, Robin Draper, and Mark Angelo. Also welcomed onboard was Sharon Chow from the Sierra Club, and Dick Gathercole from municipal politics. Many hardcore locals were sporting their Chaplin-designed two-tone blue lapel buttons—SOS: Save Our Stikine.

From the airline's flight technical staff, John Duck, who had been responsible for putting our whimsical dream on paper, was happily onboard to see how we dealt with the straight lines on his flight plan. Good luck, John. Unfortunately, Captain Bill Murray, who had donated numerous Stikine photo flights with his very special bird CF-VSB, was unable to join us; similarly, our *Save the Stikine* audiovisual maestro Blair Shakell was sadly missing. Although also invited, national icon David Suzuki and provincial spark plug Ric Careless both had other things on their plates. Most notably missing in action was the media: not one of the newsprint reporters who had been offered a free seat came along for the ride.

Onboard our aircraft, tail #720, *Empress of Montreal,* CSD Chambers and her gracious cabin crew helped everyone find appropriate seating while offering flutes of morning glory and the morning newspaper—the latest newsletter from Friends of the Stikine—providing a map of the destination watershed along with some historical background and brief descriptions of its most prominent features. Wishing you a pleasant flight!

Up front, the buzz was palpable … feeling that extra tingle of excitement invoked by first flights and check rides. In the right seat, my more senior and more seasoned first officer for the day, Captain Garry Grant, seemed pleasantly amused: uncertain perhaps about what he had volunteered for, he managed to infuse a high degree of professionalism and a trace of cockpit discipline, and with a smile on his face. Everything was ticking along smoothly until a call on VHF from Ramp Ops advised our push back would be delayed.

Apparently, our non-scheduled flight was seen as the perfect opportunity for introducing a brand-new nightshift load agent to the weight and balance form—multiple pages of lined graphs and columned numbers—to determine the loaded gross weight of the aircraft and the position of its centre of gravity as influenced by fuel and passenger loading. In conjunction with wind, temperature, and air pressure level, these numbers are used to calculate power settings and airspeeds for the take-off procedure. An inward smile in remembrance of personal struggles with the dreaded W&B form was accompanied by subtle concern. We had a narrow window of opportunity in which to work: the airplane needed to be back in time for the morning sked; the longer we sat here, the more time was available for an extremely competent ground crew to notice something amiss down below; and the more time was available for an ever-conscientious fleet manager upstairs to suddenly realize a more productive use for our long-range fuel tank. *Breathe Peter, breathe.*

We were off the ground about thirty minutes later than planned with about 120,000 pounds of airplane climbing on a standard routing directly towards YYD Smithers in a starlit sky, the normal first leg of the inland route toward Whitehorse. With Second Officer Copeland in the centre jump seat, it was a typically quiet and dimly lit cockpit: soft voices surfing over the hum of lights and instruments consuming electricity … a finely tuned machine climbing effortlessly into a timeless void enroute to our destination of choice. Behind the never-locked cockpit door, the atmosphere was distinctly different: bright lights and busyness accompanied champagne breakfasts being served in a chatterbox of anticipation.

Singing and Soaring

The timing could not have been better. Soon after checking in with Edmonton Area Control Centre at BOWSR intersection, 169 nautical miles northwest of Smithers, we obtained clearance to descend below controlled airspace (similar to an aircraft inbound to Dease Lake), and immediately overhead Mount Edziza's volcanic cone in the centre of the watershed, we pulled the plug (closed the throttles completely) to begin a slow descending turn out of FL310 (31,000 feet above sea level). Our eclectic band of adventure seekers from hither and yon were not disappointed. The day's initial glimmer of sunlight over the northeastern horizon turned the snow-topped mountains beneath us into a broad blanket of rosy-coloured gloss.

Of interest to the pilot flying was the brief time required for the area's high ground to come up and meet him. There was no choice but to skim across Spatsizi Plateau as quietly as possible without hitting any caribou. Slowing to 210 knots (250 miles per hour) over Tuaton Lake at the headwaters, aircraft control was handed to Captain Grant for the flight of a lifetime. In severe clear and flat calm, he flew us downriver at a very safe several hundred feet above the ground—the exact number escapes me—conducting lazy S-turns back and forth across the river's course, providing eager spectators on both sides of the cabin with views up and down river. All the while, his trusty co-pilot gave running commentary over the PA system, including too much of what you have already read about. Along with landform descriptions and historical anecdotes came a desperate plea for ecological sanity, especially in regard to hydroelectric megaprojects.

"TERRAIN! TERRAIN! PULL UP! PULL UP!" Suddenly, without warning (as if it could be otherwise) our ground proximity warning system (GPWS) began shouting through the cockpit's overhead speakers. It seems we were descending too quickly and/or rapidly approaching high ground out front. "TERRAIN! TERRAIN! PULL UP! PULL UP!" Never having had to deal with the GPWS outside of the flight simulator, neither of the well qualified flight-deck commanders on board remembered how to immediately silence the screaming monster. Second Officer Copeland, our ever-alert pilot not flying, limited our embarrassment by rapidly performing non-memory items from the appropriate emergency procedures checklist ... muffling the sound with two small passenger convenience cushions held tightly against the speakers ... cancelling the bell, so to speak, while the high-price help figured out how to silence the voice of inconvenient truth.

Soon after settling into our upper river pattern and patter, Garry reported a VHF transmission from an overflying aircraft (methinks Flying Tigers) requesting to relay our estimate for LVD Level Island, near Wrangell on the coast.

Anchorage Center ATC was looking for some form of heads-up before we popped up on his busy-sector scope when rejoining controlled airspace for the route home. Shoulders were shrugged on both sides of our flight deck. We hadn't looked at our computerized flight plan since passing BOWSR and we foresaw no need for it again until reaching LVD. How many more S-turns might we do? Whaddya think? Another hour and a half or so? Yeah, let's give 'em thirty-seven past the next hour. Okay, thirty-seven it is … with a wink and a knowing smile.

In total, we danced effortlessly downriver in crystal-clear conditions for two hours, banking steeply for good looks into the Grand Canyon of the Stikine with particular attention given to the principal dam locations at Site Zed and Tanzilla Gap. After gliding softly past the river's only community, Telegraph Creek, we slipped gracefully through the magnificence of coast range glaciers (pick an altitude) before checking in with Anchorage Center to join the controlled flow of air traffic along the coast. Passing Sergief Island at the river's mouth, we reached the LVD navigational fix at thirty-eight minutes past the hour, almost exactly as calculated. Never underestimate the power of a lucky guess.

We landed on Runway 08 at YVR Vancouver with first raindrops hitting our windscreen. It was 0800 PDT. Once cleaned and serviced, our airplane would be heading for SFO; after leaving our framed posters to the gate agents and grooming crews in appreciation, we headed back to the conference. The best-ever flight of my life, my all-time favourite, was over and done. If there was one that could ever be done again, that would be it. For some of us, in keeping with the policy of Canadian Pacific Airlines (CP Air) to name its aircraft in honour of significant locales, B737–200 tail-number 720, C-GCPZ will forever be *Empress of Stikine*.

By the time the flight crew returned to the International Trail and River Conference at UBC, there was excitement in the air. One of our happy customers, rumoured to be Bing Lucas of New Zealand, had apparently been energetic in a widespread call-down of all who had missed going on the flight, especially the media. "It was the flight of a lifetime and the highlight of the conference," his spiel reportedly declared. Sounded good to me. In any case, our happy customers soon infected the entire congress with a healthy dose of good cheer. In fact, from out of the crowds, it became possible to distinguish the flight-takers by their wide smiles and bright eyes. The first-ever International Congress on Trail and River Recreation concluded in good cheer with a banquet later that day. Spirits were high, but we weren't done.

Thirty-five years later, the discovery of this old-familiar navigation chart among piles of ancient paperwork brought warm memories along with a reminder of how quickly time flies. Faint red pencil markings by company flight planners identify this particular chart as the one carried aboard our flightseeing adventure back when all the radio frequencies and all the airway headings were known by heart through regular use. Being old enough to have experienced the challenges of low-frequency beacons and radio range approaches with morse code in my ears, I have always had deep respect for the trail-breaking aviators who pioneered our navigation methods: their decades of low-level operation in lousy weather were seldom easy and often dangerous. While thinking it couldn't get any better than our new-age very-high-frequency navigation system, some of us were more than surprised by the sudden redundancy of the entire chart— airways and VHF navigation aids being largely supplanted by GPS and Satnav (Global Positioning System and Satellite Navigation) technologies which now allow point-to-point routings almost everywhere. Thanks for being here.

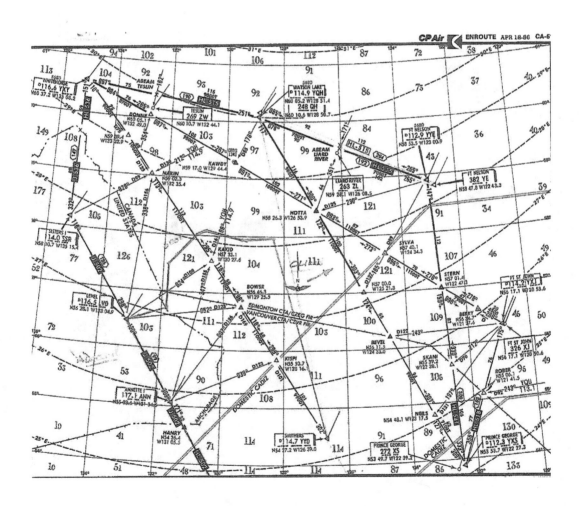

GRAND CANYON OF THE STIKINE

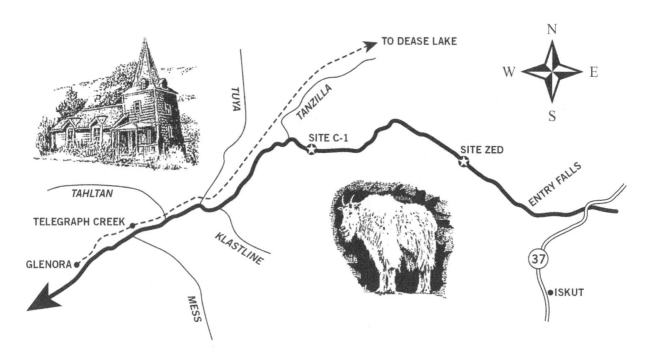

8
Stretching and Striding

A 1986 invitation for a father and son raft trip on Spatsizi River by Iskut Trail and River Adventures was impossible to resist. Working together in customary fashion, Jim Bourquin and his life-partner, Erma Nole, had organized a three-raft expedition for local young people to explore the possibility of establishing a much-needed Outward Bound outdoor education program for youth in the region. Grant Copeland and his son, Ryan, as well as me and my son, David, were offered the chance to join in. Off to the north country we went.

As any Scout masters would have predicted, heading into the outdoors with a herd of teenagers is always an education, and this trip was no exception. Fortunately, our leaders' familiarity with their charges and with the route to be travelled offset any serious concerns about risk and wilderness area vulnerability. From the village of Iskut on highway 37, a school bus for people and a stake-truck for gear took us south and then east via the Ealue Lake Road to the far-from-abandoned rail grade on which we followed the Klappan and Little Klappan rivers upstream, south, and east into the broad valley south of Spatsizi Plateau. For many reasons, this is a special place for many people: prime calving grounds for caribou and traditional hunting territory for First Nations. Among geographical features on the map, the name McConachie Peak[11] marks a beginning for some airline pilots; and for social historians, nearby Mount Gunanoot[12] says much about pioneer days in British Columbia.

More importantly, we were there immersed in a remote natural region containing the source water of four major river systems. At the very top end of Little Klappan, an imperceptible height of land between adjacent marshy ponds took us to the head of Didene Creek, which flows east into Spatsizi River. In a side valley immediately south of there, headwaters of the northbound Klappan and the southbound Nass rivers share a similarly spongy divide. So too, the northbound Spatsizi and the southbound Skeena in the next adjacent mountain pass. Slightly to the east, in and around Fire

Flats, adjacent creeks and streams feed the eastbound Finlay and the westbound Stikine. Even to the unknowing eye, this pristine expanse bubbling over with fresh water seemed to have infinite value. Lo and behold! At the foot of Mount Klappan, a harsh black rectangle suddenly appeared in a lush field of green. A large pipe pouring dark liquid into a holding pond marked the site of Gulf Canada's extensive coal deposit where tests were underway to determine its quality. If deemed worthwhile, a major mining initiative planned to knock the tops off several mountains ... without risk, of course, to local wildlife or to the wild salmon because salmon do not come above the Grand Canyon of the Stikine (their stated rationale).

By the time of this 1986 raft trip, several conservation groups in Canada and the United States were actively lobbying for various forms of legislation to protect the Stikine's inherent value. One proposal put forth by Friends of the Stikine called for creation of a headwaters national park reserve to include this particular area. Abutting two provincial parks, Spatsizi to the north and Tatlatui to the east, such a reserve was envisioned to enhance the contiguous range for wildlife while protecting headwater branches of the Nass, Skeena, and Stikine Rivers—a stopgap designation pending comprehensive land-use decisions by all pertinent stakeholders. However, not one government office, Tahltan or otherwise, expressed any interest in the idea. And while industrial concerns were understandably mute on the subject, so too were local hunters and fishers, none of whom wanted "their" resources handed over to the paper-pushers or to the enviros.

Meantime, our 1986 bus ride on the rail grade took us to trailhead where a moderate hike on gentle terrain got us and our teenaged companions to the Spatsizi River where our rafts were deposited by helicopter. Inflated and excited, we set off immediately. Unlike the upper Stikine branch, the Spatsizi has a more moderate and more consistent profile: instead of light and bouncy, it is smoother and slower, long and silky, snaking gently through an endless series of S-turns. Whatever the river current lacks in excitement, is compensated for by the scenery alongside: in its tighter valley through centre of the plateau, the Spatsizi River offers close looks at a fine array of mountain peaks. Instead of long-range views of the plateau's rounded outside rim, we got inside out cross sections filled with verticality and colour. Some peaks are gem-like in themselves; and Spatsizi Mountain, with tinges of orange and rust, is indeed the Land of the Red Goat.

A look westward up Mink Creek toward Cold Fish Lake again brought to mind the amazing story of Marion and Tommy Walker who had arrived here with their horses in 1948 after an incredible nine-hundred-kilometre overland horse drive from Bella Coola with a terrifying barge ride up Takla Lake. Employing Tahltan and other local people, the Walkers operated a guided hunting service from their home base on Cold Fish Lake for twenty years. Accessed by float plane or by the traditional horse trail, the remnants of their base lodge and out cabins are now staffed by summer volunteers, providing a BC Parks presence in the heart of Spatsizi Plateau Wilderness Provincial Park.

From our drifting rafts, we caught sight of the horse trail following Spatsizi's left bank downstream to our takeout at Hyland Post—the only community referenced on a very large map. Founded in the late 1920s by brothers Hyland, the original trading post became a secondary base in Walkers' operation and has since been maintained by brothers Collingwood as part of their extensive guide-outfitting service in and around Spatsizi. Probably always a welcome site and a comfortable spot, Hyland Post remains a great place to hang out—solid log cabins overlooking the river with a southern exposure on a large, flat bench of gravelly soil with immediate access to the high-country plateau. A crop of hay for the horses and an elevated anti-bear food cache for the humans highlight the hospitality factor; a fine little airstrip still congers up fantasies for some.

A two-night respite here allowed many of us to scramble up onto the plateau—two hours and 600 metres of easy climbing for two hours of unbelievably smooth rambling at 1,800 metres above sea level. At the eastern end of a long peninsula, we could see our river's confluence with the Stikine branch below. Looking beyond, we could make out Chukachida Corner far to the east; looking south beyond Hyland Post, we could see far up the valleys of the Ross and Dawson rivers toward their high-country origins. That day's hike gave us a small taste of a greater magnificence.

After completing our raft survey of the upper Spatsizi–Stikine, Grant and I left our sons (aged fourteen) with friends at Telegraph Creek while signing on to another adventure. Having continuously monitored the river for many years, local rafters Jim Bourquin of Iskut and Dan Pakula of Telegraph Creek determined water levels to be low enough to permit rafting the canyon section between Tanzilla Gap and Tahltan Flats. Paddled several times by expert kayakers, this section of river had yet to see a raft or any other oar-powered vessel: the previous year's high-end rafters (see below: *Hell and High Water*) had pulled out immediately after filming their on-edge passage through the gap in order to accomplish their second and more successful attempt at Entry Falls. Thus, we two paying-customers—southern fools in a northern game, according to Valdy—became part of a first-ever oar-powered technical descent of one portion of the Grand Canyon of the Stikine. Sounds impressive, eh?

Joined by photographer Fiegehen, we began our adventure at Pleasant Camp, a hunting base for locals off an obscure side road on a high bench above the canyon—a most pleasant place to camp. Next morning's hike down to the river was also pleasant—steep but manageable grassland meadows between vertical rock outcrops, taking us to the first gravel bar below Tanzilla Gap, on river-right across from King Salmon Creek, where we met an inbound helicopter slinging a net-full of river rafts and carrying our guides, Jim and Dan, accompanied by their trainee/assistant, a local Tahltan man known as Blackie. Having just completed a last-minute aerial survey of our route, they reported the river level to be still suitably low and the weather looking good. We were good to go.

After inflating our two rafts and reviewing safety procedures, we clambered onboard and hit the road. Jim and Dan were at their respective helms, manning the oars from their seat on a metal transom mounted midship while two customers on each raft were free to roam the decks and photograph at will. It was an incredible experience for all of us. The three locals finally tasted a section of river that had been tempting them for years, and we three tourists were treated to exceptional views of a seldom seen place. The expanse and variety of rock formations here are astounding: predominantly vertical canyon walls exhibiting multiple layers of columnar basalt, among other solids; cross sections of serial lava flows, varying in thickness and layered one upon another; weathered rock faces displaying crystallized artwork; deep caverns etched into layered seams; jagged spires and lava arches on scree-covered slopes; a gracefully sculpted shoreline, scoured round and black far below the high-water mark.

The river itself was a joy. Even our guides, with their considerable experience and appropriate precautions, were excitedly apprehensive about what might be around the next bend. As they had predicted, our thirty-two-kilometre journey provided two sections of serious class IV rapids within a long string of easily managed class III runs. Although not particularly life threatening for well-guided rafters in ideal conditions, this new stretch of river certainly did impress in terms of wave height and sheer power: it was the biggest river water we tourists had so far experienced firsthand. Meanwhile, interspersed sections of flat water allowed us to lollygag in appreciation of the scenery and to occasionally go ashore to observe a particular brand of mountain goats in their natural habitat—climbing down for safety—all the while under the watchful eyes of high-flying eagles. At one stop of particular interest, we explored the Tuya River confluence—a never-before-seen visual dynamic near the well-travelled vehicle bridge—large mid-channel rocks surrounded by fast-water waves and backdropped by twenty vertical feet of scoured-black cliff face below the high-water mark. Incredible.

The last sixteen kilometres of our float were equally as inspiring as the first, bouncing and sliding beneath towering walls, alert to the ever-present challenges of that day's river while gaining appreciation for its imagined high-water ferocity. Though we knew the road was up there, we saw no sign of humanity until pulling out at the Tahltan River confluence—not unlike the explorers of 1838, stumbling in where the summer rendezvous continues, albeit with far fewer participants. Fortunately, though highly charged by the excitement of exploration, we had no flags to plant nor axes to grind—all we carried was renewed respect for the river and for its people.

> ## Home
>
> Tahltans knew this place as home territory long before the modern era identified them as a segment of Athabaskan culture in the linguistic group now known as Na-Dené. Though we may never know precisely when and how they got here, we can be comfortable in saying they've been here forever. The Tahltans of the Stikine are closely related to their northern neighbours in the Dease and Liard watersheds who are identified as Kaska Dene, an organization of First Nations to the north, all of whom were often known as Nahanis by virtue of their geographical grouping. The Tahltans, by virtue of location, were known as the Trading Nahanis. To the west, the coastal Tlingit peoples are regarded as close cousins who speak a slightly different dialect; to the east, the Sekani peoples of Finlay's River are distant cousins within a farther-reaching Athabaska/Déné cultural presence; immediately to the south of the Tahltans' Stikine territory, the Nisga'a peoples of the Nass and the Gitxsan peoples of the Skeena are themselves separate entities within a distinctly different linguistic group that includes the Tsimshian peoples of the coast. To the south and east beyond all of these were millions of other people living in thousands of tribes aligned in hundreds of nations representing twenty major linguistic groups. Then came the fur traders and the gold miners and the loggers and the farmers and the tourists … and regrettably, the diseases.

At Tahltan Flats, with the riskiest section behind, we invited our boys to join with us and others for the eighteen-kilometre float to Telegraph Creek, a wonderful lower canyon experience with water big enough for a legitimate taste of the river; in fact, one tricky spot has been named Dan's Hole as a reminder of the river's ability to flip large rafts. Eight Mile Creek is the mandatory lunch stop below a twenty-metre waterfall pouring like a faucet into a nearby rocky pool. At the landing in Telegraph Creek, history came alive with the sight of once-grand homes perched hillside above the remains of a dilapidated waterfront warehouse. At the RiverSong, we were treated to another view of the canyon via videotape recording.

Hell and High Water is title of an ABC television production chronicling the technical descent of the Grand Canyon of the Stikine in 1985. Utilizing helicopters and videotape to the greatest extent possible, world-class white-water specialists took on the challenge with three kayaks and one state-of-the-art self-bailing river raft. It wasn't always pretty, but it certainly was exciting. Entry Falls stymied them right off the top and they had to go around it: portages were impossible without the helicopter and sometimes barely possible with it. Nevertheless, with determination and high-tech support, they were able to paddle most but not quite all of the canyon. Unanimous in expressing respect for the river that almost claimed one of them, expedition members avoided any reference to "man conquering nature" although their TV producers could not resist the cliché.

For paddlers of any ability, it was an education: the water was huge, and the lines were complex. Videotape technology provided an up-close, tangible look at the furious maelstroms previously seen only from far above. Noting significant changes occurring over several expeditions since his first in 1981, team leader Rob Lesser surmised the canyon's powerful current to be rolling house-sized boulders far enough downstream to gradually reposition some of the more challenging paddle sites. Anyone who has heard big rocks being moved by a big-time current would have little difficulty with the concept.

Writer Monty Bassett certainly didn't have any difficulty with it. At the time of our 1982 canoe trip, he was executive director of the Spatsizi Association for Biological Research, which had been radio-collaring Osborn caribou in a project aiming to determine their range and population in view of the proposed hydroelectric dam complex. Among numerous watershed interests, Monty also served as a point man for the 1985 *Hell and High Water* adventure, giving him firsthand contact with the goings on. Nor does photographer Gary Fiegehen have any difficulty with the concept of house-sized boulders rolling down the river. After their brief swim together at Jewel Rapid in '82, Monty and Gary teamed up for several magazine articles, including a science-inspired project investigating fossils along the canyon floor—a project that began during the '85 *Hell and High Water* adventure when Monty discovered two interesting fossils in the canyon wall that were then dated by the University of British Columbia. One specimen was a willow ancestor geologically assessed to be a hundred thousand years old; the other was a cedar bow assessed as ten million years old—of a species found in China.

Monty's fascination was triggered by discovering, contrary to expectations, the older fossil had been found in higher strata than the younger, more recent willow fossil, embedded in strata far below. Monty reports this as the first of many clues revealing canyon strata to have been repeatedly folded over onto itself due to the Pacific plate being accreted into the North American plate. Monty's canyon story:

> After helicoptering into the middle of the canyon, our camp was sandwiched between the wall and the river on a sandy beach—about a hundred feet square peppered with rocks the size of bread loafs, a few the size of four-burner cook stoves. We assumed both were remnants of ancient times, exposed by the raging river just meters away. It was a clear night. Not a single cloud to disrupt the brilliance of the stars that danced a ballet high above, magnified by the telescopic effect of the vertical walls.
>
> Then, sometime in the night, it began to rain. We each had tents and the phenomena went only mildly recorded until suddenly there came a sound like Velcro being ripped apart and rocks started cascading down. At first, the rocks flew over camp and into the river, but suddenly one ripped through the tarp covering our kitchen and crashed into the pots and pans! It seems droplets of rain were forming rivulets and streams that undercut the tenuous strata [rock] that in turn tore away from the canyon wall. Trapped by the downpour of rain and debris, I salvaged a pot bigger than my head and wore it as a helmet beneath a turban of climbing ropes while huddled in my tent and pinned to the side of boulder hoping it would deflect a full-out mudslide. Still, the sound of tearing Velcro, then silence and then the crash of the rocks around us. "You alright?" we'd call

out back and forth, relieved with the sound of another human voice. I also remember negotiating with my ancestors for our lives.

At the first vague light of dawn, we did a quick inventory to discover our kitchen tarp was shredded by a half a dozen loafs. The decision was made to immediately consolidate down to life-support essentials, and take-off downstream to the wide delta mouth of a side stream we had seen on the flight in … two to three hours below.

The river-rocks were snot-slick with a wet-moss glaze; our packs soaked and heavy. Still, we made two grueling trips with all of our gear down to the safety of the wide tributary. But then, less than a quarter of a kilometre to go, there came a reality shattering, telltale rip of Velcro so loud that it echoed off the walls of the canyon and masked even the roar of the cascading river just feet away. Gary had gone some twenty meters ahead, while I'd stopped to tie my boots: the granitic consequence of that enormous Velcro tear exploded off of a ledge high above us in otherworldly fashion … followed by free-fall silence … before crashing against another ledge close above us in deathly fashion … followed by free-fall silence. We both dropped to our knees not daring to look up.

Bread loafs and kitchen ovens were now small potatoes: this boulder was the size of a Volkswagen Beetle, and the splash of its impact into the river threatened to sweep the Sherpas and their gear into the Stikine. A week later when the helicopter came to pick us up, we flew upriver past our original camp which was now the terminus of a giant, impounding mud slide.

"Live and learn," recalled Monty, sometime after the fact. "We should have known something was wrong when all the goats disappeared from the canyon walls." Recalling the experience, Gary told me he would never look at those canyon-wall scars in the same way again. With innate respect further enhanced, these human survivors have continued on their ways with great interest and concern for the river. Among a series of publications about First Nations and their lands, Gary has produced a fine coffee-table book: *Stikine: The Great River*. Also, seldom idle, Monty has produced entertaining and informative films about the Stikine and its critters. Among others, see *Life on the Vertical* and *Written in Stone*.

My own creative instincts kicked into gear about this time with Gary's generous gift of several hundred duplicate Stikine slides, which were sorted and edited down in sync with my music of choice and arranged accordingly into one large carousel for a slide projector. On the scheduled evening, in the Planetarium Star Theatre of H. R. MacMillan Space Centre in Vanier Park (circumstances long forgotten), an audience of considerable size was treated to numerous large images from a wildly natural place few had ever seen, all being displayed in quick time to a musical soundtrack through a quality sound system. Thanks go to Mark Knopfler and Dire Straits for their wonderful "Telegraph Road" composition that so aptly accompanied looks at our own dirty old track.

HISTORY 202

Subsequent to Samuel Black's upper Stikine exploration of 1824, the Hudson's Bay Company (HBC) continued expanding its territory north down the Mackenzie River and west up the Liard River to counter Russian fur-trading activities on the Pacific Coast. In 1838, HBC trader Robert Campbell found himself with a challenging itinerary. From Fort Halkett (at Smith River, about 180 kilometres upstream from the Mackenzie), his party of eight men and two canoes battled a further 160 kilometres of dangerously high water upstream on the Liard River mainstem before paddling (and pulling) another 240 kilometres up the Dease River, a southern tributary named for their honoured senior, Peter Warren Dease.

While half of Campbell's party were assigned to construction of a trading post at its Dease Lake headwaters, Campbell and three trusted colleagues ventured into unexplored territory, southwest beyond the Arctic–Pacific divide, looking to reach a certain westerly flowing river rumoured to be rife with trading activity. Their timing could not have been better. A one-hundred-kilometre trek along the age-old animal-hunting-trading trail brought them to Tahltan Flats at the confluence of a mighty river and a formidable tributary where Campbell reported the largest gathering of Indigenous people he had ever seen. Fortunately for them, the HBC men had initiated friendly trade agreements and shared a peace pipe with local Tahltans they had met on the inbound trail. Otherwise, their welcome might have been short-lived—literally.

It was the summer rendezvous, an annual meet-and-greet trading symposium centred on unlimited numbers of fresh salmon being harvested by hundreds of people from far and wide. Activity was at its peak when these strangers stumbled in from the east and unintentionally created alarm for visitors from the west. The Tlingit people who live at the mouth of the Stikine on the Pacific Coast were the customary A-list attendees at this annual event: for generations unknown, they had been sailing, poling, paddling, and pulling their canoes upstream for 280 kilometres, at least once per year, on what is now recognized as the continent's fastest-flowing navigable river, to talk and to trade, and to meet and to marry, with their inland cousins. The Tlingit brought sea otter pelts and shell-ware to trade with the Tahltan for caribou hides and obsidian.

Migrating salmon established the timing, fresh salmon fed the masses, and dried salmon was everyone's food for the future. Also coming upriver with the Tlingit were trade goods in the form of *modern* manufactured items recently acquired from Russian traders who had become established on the coast. In fact, the Tlingit had become very protective of their preferential trading status, up the river as well as on the coast. For some years now, the Tlingit chiefs had been advising the inland Tahltans to kill any white man approaching from the east. Given that several Russian agents had also travelled upriver on this particular occasion, the immediate future looked pretty grim for Robert Campbell and his colleagues.

Fortunately (again) for these uninvited guests, the female chief of the local Tahltans was more into hosting than hostility—she contributed to the Campbell party's protection and facilitated their escape. On his way out of town, Campbell is reported to have planted the company flag and engraved a tree, taking possession of the country for the Hudson's Bay Company. Very soon thereafter, saved by food donations from their new Tahltan friends, these company men barely survived a dreadful winter at their Dease Lake dwelling, eating the boiled rawhide from their snowshoes before paddling quickly down to Fort Halkett on the Liard. Meanwhile, the trade-protectionist Tlingits of that 1838 rendezvous were probably the most surprised of all when, in the very next year, their homeland on the coast was leased away by the Russians … to the English.

Never venturing back into Stikine country, Robert Campbell later explored the upper Liard River where he established a trading post at Frances Lake before becoming the first European from the east to enter the Yukon watershed where he established Fort Selkirk. His name is remembered on our maps by a mountain peak and a highway. Subsequently serving as a chief trader at several inland districts until 1871, Robert Campbell died in 1894 at his Merchiston Ranch, near Riding Mountain in Manitoba. [13/14/15]

9
Dashing and Dancing

FOS newsletter #15, which was entitled OUR COMMON FUTURE, was another masterful piece of work by Grant Copeland and colleagues. Acknowledging Environment Week 1988, the newspaper-style spreadsheet featured a front-page photograph by Gary Fiegehen, which captured two boys of Telegraph Creek resting astride their bicycles side by side, both wearing facial evidence of a raid on the choke cherry bushes behind Dora Williams' house near the RiverSong. While the bikes and clothing of the two young lads were quite similar, their baseball cap logos were appropriately dissimilar: Andrew Fisher's advertised the Calgary Stampede and Steve Marion's celebrated Tahltan Country. My sincere thanks to FOS director Carol Lambert whose relentless promotion encouraged the selection of this cover image, which also inspired a poster.

As well as providing an update on our watershed concerns, that current issue of the newsletter from Friends of the Stikine outlined provincial land redesignations arising from a recent busy period of land-use analysis. While industrial development was banned in four of its major Class A provincial parks and adjacent recreation areas, the parks system was augmented by a land area greater than Prince Edward Island; most of the additions were designated as recreation areas and scenic corridors with minimal restrictions on resource extraction. Nevertheless, steps were being taken in a decidedly different direction, for which some observers credit significant voices from the outside world.

Toward Sustainability

In June 1987, the United Nation's World Commission on Environment and Development (WCED) released its landmark report, *Our Common Future*, which thoroughly documented our failure to effectively integrate social and environmental considerations into the process of economic development. The twenty-two members of the WCED called for unprecedented levels of international cooperation and public participation in the decision-making process in order to properly manage our global environment. Otherwise, as stated by Madam Gro Harlem Brundtland, prime minister of Norway, who headed the commission, "If we do not succeed in putting our message of urgency through to today's parents and decision makers, we risk undermining our children's fundamental right to a healthy, life-enhancing environment."

To combat overwhelming threats to our survival as a species, the report also called for a forty-per-cent reduction in energy consumption by developed countries of the world while recognizing conservation, not more dams, as the lowest-cost alternative for making more energy available. Alongside many aesthetic reasons for conservation, the economic value of genetic materials alone is sufficient to justifying species preservation—commercial medicines and drugs with their origins in wild organisms had at that time, an estimated world-wide value exceeding $40 billion *per year*. The commission recommended the total expanse of protected areas needed to be at least tripled if it was to constitute a representative of Earth's ecosystems. Protected areas in North America were then considered to represent 8.1 per cent of total land area; at the time, protected areas in British Columbia free of industrial development were calculated at five per cent of the land base.

The WCED report also advocated recognition and protection of Indigenous peoples' traditional rights to land and other resources that sustain their ways of life. "Disappearance of these people is a loss for the larger society which could learn a great deal from their traditional skills in sustainably managing very complex ecological systems."

The Canadian response to this report from the World Commission included establishment of a National Task Force on Environment and Economy, calling for the implementation of sustainable development strategies for each jurisdiction by way of broadly based consultative groups at all levels of government (round tables), bringing together leaders of all sectors to foster ideas on sustainable development in Canada, and to provide leadership in putting those ideas into practice. From the Honourable Tom McMillan, Canada's minister of environment, speaking to the United Nations General Assembly in New York on October 19, 1987: "Mere tinkering with the status quo is a prescription for failure. What is required, in particular, is a change in the way people think—the most challenging change of all. To my mind, the relationship of Canada's native peoples with the natural environment provides a model in sustainable development."

Our Telegraph Creek Convention of 1985 had recommended all land-use proposals within the Stikine Watershed be subject to settlement of the Tahltan Aboriginal Land Claim, and that Tahltan people be employed in management of any protected area designation within their territory.

Also in 1987, the Fourth World Wilderness Congress (not trails and rivers) at Estes Park, Colorado, attended by Corinne Scott, John Christian, and Grant Copeland of FOS, enthusiastically endorsed a resolution urging the governments of Canada and the United States, British Columbia, and Alaska to establish a Stikine Transnational Park to include the entire mainstem of the Stikine River while also calling on the states of Washington, Oregon, and California to reconsider the construction of transmission lines and the purchase of energy from British Columbia. The Washington Wilderness Coalition, a co-sponsor of this resolution, declared in the words of Executive Director Pam McPeek, "Studies prove that conservation is a much more cost efficient and environmentally benign alternative to meet future energy demand. There is no need to dam more BC wilderness rivers to provide electrical energy to the United States."

The U.S. National Parks and Conservation Association, together with Sierra Clubs in Canada, supported proposals for a Stikine International Park and for a Stikine Biosphere Reserve as well as for a Spatsizi World Heritage Site. As stated by Edgar Wayburn, vice president of the Sierra Club, "The magnificence of this river [is] overwhelming in its quality and diversity. Such a wilderness treasure should not be lost."

At its 1988 meeting in Costa Rica, the International Union for Conservation of Nature (IUCN), the world's largest non-government organization concerned with conservation and preservation issues, put forth a resolution recognizing the immediate need to protect the Stikine as an international park.

As summarized in *Our Common Future* newsletter of 1988, the Stikine Biosphere Reserve proposal, though still on the table, had gathered little momentum other than the IUCN's documented support for the transnational park recommendation. Regardless of any biosphere reserve status, FOS suggested creation of Tahltan managed youth-oriented rediscovery programs to monitor wildlife populations without seriously impacting the resource base.

Given the degree of international recognition accorded Spatsizi Plateau, its wildlife populations, and its related source waters, a Spatsizi World Heritage Site was proposed to UNESCO (United Nations Educational, Scientific, and Cultural Organization) as a valuable addition to the list of the world's outstanding cultural and natural wonders. Canadian sites already listed: Kluane National Park Reserve (YT), Anthony Island Provincial Park (BC), Nahanni National Park Reserve (NT), Wood Buffalo National Park (AB).

Given the unknowns of international recognition for the watershed and protection for the main river corridor, there remained room for the headwaters national park reserve proposal, which also evolved from the Telegraph Creek Convention. Noting the Spatsizi River to be nearly one-half of the Stikine's upper drainage and noting its source to be outside the provincial park, a headwaters reserve (a personal favourite) would straddle the Stikine's southeastern divide to abut Spatsizi and Tatlatui provincial parks, while including the unprotected upper reaches of the Spatsizi, Klappan, Nass, and Skeena Rivers. In providing a suitably large contiguous area for wildlife populations, this protected area would also add a valuable link in the Canadian national parks system with a superb example of the high plateau country of Natural Region #7.

Also remaining on the table (centremost) was Gulf Canada's Mount Klappan coal project, which had avoided any reference by the recent Wilderness Advisory Committee, despite washing units and holding ponds sitting beside the Spatsizi River at the very head of the Stikine Watershed. Gulf Canada's prospectus, which remained the basis for all environmental study, assured there was no threat to the fishery because salmon did not get above the Grand Canyon of the Stikine. Brilliant. Facing a multitude of requests for a public hearing prior to approval of a seemingly invisible three-year-old "test dig," the province's Environment and Land Use Committee was deciding not to decide anything just yet. Given the poor optics of Crowsnest and Tumbler,[16] it was difficult to believe the BC taxpayer could be expected to underwrite yet one more of the same, especially one situated on the fragile divide among headwaters of three major river systems, on the high plateau as yet unrepresented in the Canadian national parks system, in prime calving grounds for Osborn caribou ostensibly protected by the adjacent Spatsizi Class A Wilderness Park.

As envisioned at the 1985 gathering in Telegraph Creek and described in Grant's comprehensive visitor's guide, the Stikine National Park Reserve proposal received overwhelming support at that year's assembly in Banff on the occasion of the one hundredth anniversary of the Canadian national parks system. Although a national park does not necessarily follow, the process-oriented reserve designation temporarily suspends resource exploitation while land claims and other broad issues are addressed. Any park resulting from this national park reserve process is committed to inclusion of Indigenous peoples in park management. To date, there had been no apparent action on the proposal.

The Canadian Heritage River proposal was also still out there. A perennial favourite of many river lovers in British Columbia who preferred their politicians to look at something other than the sale of cheap power to California, the heritage river option remained academic because of the provincial government's (thus far) refusal to participate in CHRS.

Hope remained, however. In March 1987, concurrent with implementation of a lower-river Forestry-managed corridor, the BC Ministry of Environment and Parks (under new minister Bruce Strachan) announced creation of a 217,000-hectare (536,000-acre) Stikine River Recreation Area along both sides of the river upstream from Telegraph Creek and encompassing the Grand Canyon of the Stikine while abutting Mount Edziza and Spatsizi provincial parks. Purportedly in recognition of inherent recreation values of the upper river, this particular designation still allowed for the possibility of resource extraction and hydroelectric development within its boundaries. Reportedly years away from implementation of a management plan, the Parks Branch began acquiring existing trap lines while upholding BC Hydro's *right* to a flooding reserve above the canyon. The level of regard given, if any, to the Tahltan land claim and/or to the traditional *rights* relationship between trap lines and flooding reserves was unknown.

Then Again

The road ahead was still looking a bit congested. On December 13, 1986, an order-in-council of the British Columbia cabinet had allocated the entire Cassiar Timber Supply Area for wholesale log export. A pre-election freeze of raw log export had disappeared without discussion or public input. The forthcoming five-year plan (yet another) was seen to concentrate on easily accessible timber close to the highway, in the heart of the watershed between Spatsizi and Edziza. We feared that this logging boom could lead into the upper Iskut Valley, which in turn could lead the way to clearcutting of the Klappan Valley and mid-Stikine in and around BC Hydro's future reservoir. This scenario did not seem ridiculous to anyone who had seen what was already happening along Highway 37—one of  the province's most scenic highways had been turned into a haul road for a massive and seemingly desperate tree-mining operation. The district Forestry Office was feeding local inhabitants with rumours of an impending moratorium while hundreds of trucks per day were rolling into Stewart with whole logs for export. Not holding our breath in hopes of a moratorium, we were expecting to see the formation of yet another committee and a concurrent increase in the annual allowable cut.

Things were still happening on the lower river as well. Despite two public studies on the proposal, the BC Ministry of Forests ignored the widespread opposition and had proceeded with a "test cut" several years ago; logging continued with an airstrip, base camp, and heavy equipment on site. While most of the economic gain flowed beyond BC's borders to China, which was buying the logs, and to the US for supplies, shipping, and handling, British Columbians were bearing the cost of clear-cut logging on the "scenic" lower Stikine.

As such, a process aiming toward creation of the Lower Stikine Recreation Corridor that began in 1987 must have been complex as well as lengthy. Perhaps the financial returns were insufficient for the first two logging operations permitted in the lower river and/or the Tahltan interests were badgering to get in on the action. Nevertheless, as Mark Hume reported in the *Vancouver Sun* (February 9, 1988): "In a surprise move, the provincial ministry of forests has endorsed wilderness preservation on the lower Stikine River ... The forest service plans to ask the ministries of environment and mines to support its proposal for there to be no logging or road building in a recreation corridor along the Stikine River from the Alaska Panhandle to Telegraph Creek."

To be clear, this was not wilderness designation per se—already rejected because of the river's high volume of motorized boat traffic. This was simply the protection of wilderness values within a *recreation area*—a designation intended to permit logging under approved guidelines. Nevertheless, citing public opposition to logging and recent international expressions of concern, Jim Snetsinger, regional recreation officer for the forest service, stated: "... no large-scale logging permitted along the lower Stikine ... We are willing to forgo any long-term logging interests in recognition of the scenic and recreational values."

Beneath the ever-present shadow of hydroelectric dams upstream, this first-of-a-kind corridor was seen as a positive step forward ... for the moment. However, it was never put in place; or, if it was, it disappeared again with nary a trace. After a year of blitzing the media and NGOs with Forestry's new enlightened view of wilderness values and recreation potentials, this lower-river proposal suddenly dropped off the radar. Although particular reasons are lost to history, it seems highly possible that, when all the chips were down, the responsible agency, MOF, found too many competing interests to allow for successful creation of this first-of-its-kind land-use designation. The whole idea was tabled for further study. A more cynical voice might suggest this sudden and temporary spate of enlightened forestry was but a ruse to deflect attention away from the humungous clear-out sale along Highway 37.

Battle for Resources

Out in the bigger world, the fish wars were still raging. In 1987, fisheries officials announced that Canadian fishers would harvest forty-two per cent of the Stikine run despite a lack of agreement with the US government. According to Mark Hume, writing in the *Vancouver Sun* (July 7, 1987), the Canadian position was taken after negotiations failed to increase the existing quota, which had only allowed Canada thirty-five per cent of the Stikine run. Estimating that at least ninety-five per cent of the transborder salmon spawn in Canadian waters, Gordon Zealand, DFO supervisor for the area, said Canada's share of the transboundary catch was not realistically reflected in catch allocations. Steve Pennoyer, chairman of the US northern panel of the Pacific Salmon Commission called the Canadian initiative a violation of the treaty while claiming traditional rights to the salmon because Americans had been fishing the Stikine and Taku rivers for a hundred years, and Canadians didn't start fishing there until 1979.

Discussions continued. In the following year, 1988, Canada and US negotiators achieved a major breakthrough with a deal that overcame the deadlock of almost ten years. Although both sides had to give a little, the Canadian share of the catch increased immediately and was planned for further increases if agreed-upon enhancement projects proved successful. Expecting to raise the Stikine annual sockeye run from thirty-five thousand to one hundred thousand fish, the enhancement agreement called for eggs, three to six million per year, to be milked at Tahltan Lake in BC and delivered by air to Snettisham Hatchery in southeast Alaska for incubation, after which the fry would be planted back in Tahltan Lake and nearby Tuya Lake.

> Meantime, Jim Fulton, member of parliament for Skeena was decrying the lack of manpower (only two game wardens in northern BC) and lack of action to address reports of large-scale poaching of moose from the lower Stikine by illegal entrants from Alaska.

Farther upstream, after failed negotiations, a group of Tahltans in Telegraph Creek seized three Caterpillar tractors, two Kenworth trucks, a John Deere backhoe, fuel tanks, a mobile bunkhouse, and a mobile kitchen—with total estimated value of $1 million—while it was in transit to Golden

Bear Mine, 140 kilometres to the northwest. An immediate settlement of the dispute saw the Tahltans be assured a minimum of twenty jobs at the mine, owned by Chevron Minerals Canada and North American Metals, as well as receiving guarantees for maintenance work on the road along with several other major concessions. "It's probably the best deal any Indian group ever got out of a mining company," said Jerry Asp, president of the Tahltan Nation Development Corporation.

Meanwhile, over on the Iskut River, Skyline Explorations poured the first gold ingot from its Johnny Mountain Mine, culminating eight years of exploration and development by Reg Davis and company. A *Wrangell Sentinel* feature of August 25, 1988, described the scene as a champagne-sipping celebration for about 150 fly-in guests who got a look at the mine's extensive infrastructure at the 1,200-metre level, "where the trees don't grow and the ground is often white with snow." Without an access road, and therefore fully dependent upon air access, Johnny Mountain Mine was a remarkable achievement; and with some 12,142 hectares (30,000 acres) staked in the neighbourhood, Skyline was not done yet.

Mining activity in general was already heating up on the ring of fire. In January of '89, Cominco announced an expenditure of $50 million for construction of a full-scale gold mill and camp at its sixty-per-cent-owned Snip gold and silver property on Snippaker Mountain (adjacent to Johnny Mountain). The other forty per cent ownership of this property was in the hands of Prime Resources, a multi-merger entity arising out of Vancouverite Murray Pezim's numerous mineral stock-flogging forays. Pezim said the mine would create three thousand jobs and operate for ten to fifteen years. So far, workers from Vancouver were being flown via Smithers into nearby Bronson Creek airstrip, while bulk equipment was being barged to Wrangell for subsequent shuttle by river and/or air to the mine site. Ominous.

On the Coast Mountain Divide very nearby, Alaskans were making plans for construction of a transmission line from their Tyhee Lake generation site to supply electricity to the new crop of mines on the Canadian side. In considering the prospect of a deep-water port on adjacent Bradfield Canal, the citizens of Wrangell and Petersburg were wrangling over the division of profits from the sale of this electricity. Farther south, the proposal for a grizzly bear sanctuary in the Khutzeymateen Valley was receiving enthusiastic support from Prince Philip, Duke of Edinburgh, and president of World Wildlife Fund. In the outside political world, Maggie Thatcher and Ronnie Reagan were hard at work making their economies supposedly more sustainable.

Back home, in 1988, BC Hydro established a wholly owned subsidiary known as Powerex to be the exclusive exporter of BC electricity; and while this semi-separate entity was catering to foreign and Canadian companies lining up to generate the stuff, "clean-green" BC Hydro was advocating more efficient consumption of electricity in order to stabilize its costs and to also stabilize the rates charged to customers. Oh, really? How considerate. Despite threatening rhetoric and questionable energy forecasts from BC Hydro, the voters of British Columbia seemed to hold sway, at least in the short term. An upwelling of public awareness in tune with emerging technologies such as wind, solar, and geothermal was sufficient to see the Stikine–Iskut five-dam megaproject put on the back burner for an indefinite period of time, finances being reported as the major factor.

Although we often seem to have progressed little in a thirty-two-year span since the *Our Common Future* heads-up was issued by Madam Brundtland, the Stikine has fared much better than many other rivers on planet earth. Introduced quietly in 1987, the Stikine River Recreation Area on the upper river appears to have served its purpose well. Miners and dam builders were comfortable in recognizing scenic and recreation values as long as their resources remained available. Appreciating the recreation area as a possible step forward, the conservation community was able to tone down its campaign rhetoric while keeping a close eye on rampant logging operations for any hint to the direction of BC Hydro. First Nations of the Stikine and the Province of British Columbia did reach agreement on trap lines and other traditional activities—probably not an easy process. In due course, the mining industry was appeased by offers of alternate access routes and/or substitute properties. BC Hydro gradually retreated from the canyon and its Stikine–Iskut five-dam megaproject soon disappeared from the newspapers. Then followed a decade of silence on the matter.

Finally, in 2001, the upper-river recreation area was upgraded to Stikine River Provincial Park status, thereby installing another layer of anti-dam protection on the Grand Canyon of the Stikine. While some tributaries continue to support small-scale hydroelectric installations, the mainstem Stikine so far remains 640 kilometres of wildly beautiful river in a near-natural state.[17]

10
Resting and Reflecting

My story continues into a 1989 upper-river adventure with my father Edward (Ed) who had long been listening to my rantings and ravings, and who had been happily, and sympathetically aboard the *Empress of Stikine* sightseeing charter flight in 1986. Although we had enjoyed ample experiences together in the backcountry, this trip was an opportunity for me to give something back to my old man while sharing time in a special part of the world … the canoe trip he couldn't give me when I didn't know I needed it.

Fulfilment came three decades later when Murray Wood and his beloved Beaver flew father and son, along with canoe *Dimples,* to a familiar campsite at the southeast corner of Tuaton Lake where a morning of moose in the shallows and fish on the line aptly introduced the wilderness experience. Words can barely express the amount of joy found in sharing this magnificence: sparkling lakes and natural beauty undisturbed by noisy humans (other than ourselves), or by threatening weather. Later, while camped in warm light at the foot of Fountain Rapids, a lifetime of fond memories returned with the sight of Daddy Ed doing his bannock and beans over an open fire.

From there, the trip around the southeast corner of Spatsizi Plateau was everything expected and more. In ideal weather and water level, we boogied almost nonstop, running riffles, and bouncing around boulders with great abandon and obvious good cheer, all the way to Adoogacho Creek—a forty-kilometre day that had included a portage at Chapea Canyon. Pulling us onto the beach at destination, Dad collapsed backwards and remained motionless. Hypothermia! Although he was having too much fun to complain about cold water splashing over his bow gunwales, the last hour and a half in that evening's rapidly cooling air had been a bit much for his lightly clad body … and also a reminder for me to err on the warm side whenever in doubt. Fortunately, this situation was quickly resolved by my 1985 kindling stash found untouched beneath the nearby spruce tree, and by my father's previously undeclared mickey of whisky theretofore being saved for my forthcoming birthday celebration. Close enough. Medicine is medicine.

The following day was a long one, seeing us head uphill with minimum gear, stopping for photo ops and appreciative views at top of the falls before bushwhacking up the creek (on the route first seen in 1985) to make camp at the treeline in a steep meadow filled with wildflowers. Encouraged by continuing fine weather, we pursued our (my) primary agenda item—a day of exploration on an alpine plateau. It was wonderful. With sufficient wind to keep the bugs down, Dad wore a big grin beneath his bandana-bound head while wandering uncluttered terrain on the westernmost spur of the map's Claw Mountain. He loved it. The sixty-eight-year-old mail carrier was easily up to the task, and my respect for him increased appropriately. The journey had proven its worth. While exploring ridgetop views, we came to appreciate the relative softness of Spatsizi Plateau in defined contrast to the rugged topography of its surrounding mountain ranges which challenge all who dare. Alongside ancient charms on the latest topographical map we found reminders of more recent history:

Thudaka   Kechika   Chukachida   Omineca
Samuel Black   Albert Dease   Finlay and Swannell

After a second night up top followed by a birthday-bash evening on the beach, we did another forty-kilometre day that got us into Sanabar Creek where a highly abbreviated version of my *s-a-n-d-b-a-r theory* joined a host of stories told around a heavenly campfire. Next morning was hell. The sound of determined raindrops on nylon prompted sudden awareness and a call to action. So much for erring on the warm side. A week of clear, dry weather had combined with laziness and fatigue to omit installation of the tent's rain fly once again … its first line of defence. Dumb. Methinks we were hit by a fast-moving cold front, and some guys never learn. *Too soon old and too late smart.* Dad's age-old admonishment hung in the air, unspoken. With help from a panic-rigged fly, we kept sleeping bags dry while managing hot coffee by our small gas stove before packing up a very wet tent. Ugh! This trip in '89 was my first and last time to ever be caught with my tent fly down.

Having survived rude awakening, father and son idled past the mouth of River Spatsizi while sensing drier and cooler air behind our surmised cold front. In no particular hurry, a few kilometres downstream, we stopped to investigate Holmes' Hunting Camp, the modest assortment of log structures and corrals on river-right. Surprise! Unlike my three previous visits, somebody was home. Similarly surprised by finding humans at their front door, caretakers Joy and Ben Guenter of Smithers warmly entertained us for two hours with fresh coffee and lively conversation accompanied by the world's best chocolate-iced cupcakes—unexpected bounty from the only extracurricular people encountered during multiple trips on the upper river in a fifteen-year span.

Immediately downriver came Jewel Canyon, its set of rapids approached in lower water conditions and with a higher level of awareness than on two previous occasions. Several lapfuls of fresh water for my bow-mate saw us through in lively manner. Ever-lamenting loss of the once-gorgeous campsite at the Pitman confluence, long obliterated by powerful river dynamics, we pressed on in search of distance under a clearing sky and decreasing air temperature. After getting as much daylight as possible out of a healthy current, we took out on a river-left gravel beach piled high with weathered driftwood, and while our sodden tent received the drying warmth of a candle lantern, father and son spent the night under an infinite blanket of stars while basking in the warmth of a near-infinite supply of firewood. Perhaps because of Holmes' camp coffee, sleep didn't seem necessary for me. Absorbed by the immensity above, I became entertained by the minutiae below—rattles of gravel sliding downhill and muffled clunks of river-bottom rocks on the move. Ever the teacher, Stikine added another verse to "finding the greatest of pleasures in the simplest of things."

As usual, Beggerlay Canyon came up out of nowhere and proved as impressive as ever. Adjacent to the sandy-shelf campsite on river-right, a skookum new foot bridge had been installed over Beggerlay Creek as part of a portage route arising from creation of the Stikine River Recreation Area two years prior. A great aid for exploring the canyon, the portage bridge proved unnecessary for our paddling needs: once again, the old eddy-out, forward-ferry, panic-turn technique saw us safely into the chute amid whoops of joy enroute to canyon serenity—a small rush of excitement in celebration of a great journey.

In honouring his commitment to guide a seniors' hiking excursion at home in White Rock, Dad flew south out of Iskut while I ferried our gear south in my new used "Silver Dart" pickup truck. Along the way, I had ample time for reflection and reminiscence: good fortune in the opportunity and happy celebration of a family bond through shared experience; the river had helped me learn more about myself while seeing more of my father.

On the Road

The road ahead was a dusty track guarded by stately trees having wide-draping crowns silhouetted against a leaden sky. The corridor beneath was an unholy brew of white noise and purple haze where people and cows, freight trucks and limos, bicycles and cattle, three-wheeled tuk-tuks, and four-legged mongrels do-si-doed forward in opposing directions with due regard for the footing beneath. The air was rich in animal matter and diesel exhaust. Except for the very occasional tight-strung speed freak trying to beat the system with insane weaves and accelerations on the edge of life, the scene was remarkably calm, almost serene—an eclectic blend of humanity sifting along in unhurried fashion, going about daily business. This first look at a suburban road in India came in stark contrast to recent scenes along British Columbia's Highway 37 where convoys of determined logging trucks were leaving little space for anything but dust.

Speaking from her previous experience, my travelling partner and now wife, Mary Prem Gatha, had said India would change my life. It did. To start, the 150-kilometre midnight taxi ride from Bombay (Mumbai) to Pune took me to another reality while raising my fear level to newfound heights. Later, in coming to terms with the size of India's population and its less prevalent technology, I needed to re-examine many preconceptions about efficiency and productivity, industry and economy. I love cutting wood and splitting wood, stacking wood, and burning wood. I love logging and have never disparaged anyone who earns a living in that process; however, the contrasting appearance of two dusty roads located far apart—here in India and back in Cassiar District—did inspire questions around being *civilized*. Similarly, roadside residents with thatched brooms cleaning the path for pedestrians inspired another way of looking at maintenance issues—large problems solved by utilizing local resources (Indigenous peoples being responsible for resource management?). Although these financially challenged caretakers on the subcontinent of Asia would certainly welcome a coin or two, only a smile of appreciation was ever required. Namaste.

Another great contrast I found was within the Rajneesh Ashram in Pune where serenity prevailed—hundreds of rosy-clad sannyasins from everywhere in the world gathering in solemn pursuit of inner peace. Workshops and meditation groups functioned in the spirit of the movement's founder with emphasis on love and celebration, courage and creativity, as well as on that ever-important quality known as humour. Having returned to India from the United States under questionable circumstances (better left for another story), Bhagwan had become known as Osho while reigniting inner flames at his former campground—meditation and mindfulness as the means for escaping static belief systems and religious suppression. The customary morning and evening discourse sessions brought us together in community while listening to readings from the master who commented on the writings of religious traditions, mystics, and philosophers from every written culture. Renderings of such taped discourses have contributed to more than 650 books credited to Rajneesh and available in more than sixty languages through more than two hundred publishing houses. Through the gauze of history and language, common denominators are often heard, most significantly the importance of love in the social equation: whatever else we might share, we humans all share the fact of being imperfect and our journey is mostly about trial and error in a long process. Although the scenery is constantly changing, we are on the same road we have always been on. Patience and humour are vital.

Osho would die the following year, and my marriage with Mary Prem would die soon after: not necessarily related events, but it seems we had each become increasingly committed to different goals. No regrets. The Indian Ocean sunset was as unforgettable as everyone said it would be, and the road out of India came with an overnight bus ride equally exciting as the inbound taxi ride. Patience and humour remain vital.

On the Desk

Back home in southern British Columbia, a message from Friends of the Stikine soon had me in touch with the Ministry of Forests in Smithers. FOS, our small-time river-advocacy group was being invited to join a multidisciplinary committee being formed to advise on management decisions for the lower Stikine River. The Lower Stikine Management Advisory Committee or LSMAC (El-smack) would be a long-term work in progress with reasonably frequent meetings to begin in Smithers. Oh, boy; here we go again. Following up on our 1988 report about the on-again-off-again recreation corridor then being touted by the MOF, this rant in our November 1989 newsletter helps (maybe) to further explain the situation:

> A major involvement for the past two years has been our ongoing participation in the Lower Stikine River Recreation Corridor Management Plan Study (LSRRCMPS?) under the auspices of the Recreation Branch of the BC Ministry of Forests in Smithers. The time and energy expenditure has been considerable. The result is far from satisfactory. Encouraged by MOF's apparent desire for an environmentally sensitive corridor, we had agreed to participate.
>
> Although the initial LSRRCMPS meeting had established a "recreation priority" for the lower Stikine (i.e., no roads, no commercial logging, and retention of its wilderness character), a series of management plan revisions quickly eroded that original commitment. What is left is a logging plan little better than the 1984 study which paved the way for the soon eroded "clearcuts-to-the-bank" at mouth of the Iskut. While recreation use is to be monitored to avoid conflict with nature of the corridor (i.e., the Integrated Resource Zone), there is no such concern to maintain low levels of industrial use which are allowed to continue under "strict" government-approved guidelines to ensure they do not adversely affect the river (i.e., the water which has already gone by).
>
> Although our participation appears to have been useful in refining the management plan *text,* none of our recommendations found their way into the *plan*. Not only have the boundaries been significantly altered, but the identified Corridor Management Units are now called Timber Management Units. ... After leaked reports for a planned "Recreation and Transportation Corridor" for the lower Stikine had been quickly stifled, the Environment and Land Use Committee of the provincial cabinet quickly approved MOF's corridor plan without it ever being available for public review and comment. No longer a "Lower Stikine Recreation Corridor Management Plan," it is now a "management plan for the lower Stikine corridor." A newer and fancier edition of the 1984 study, this one does pay lip-service to recognizable resources other than living trees. It is a logging plan, nevertheless. Although the lower Stikine will apparently

not be logged in the name of *recreation,* it might still be logged; it remains vulnerable to successive developments subject only to ever-present undefined stringent guidelines.

Having subsequently been invited to participate in the Lower Stikine Management Advisory Committee (LSMAC) now being created to review *development* proposals, Friends of the Stikine must evaluate its potential effectiveness in that body heavily weighted with resource extractive interests. Can a volunteer group of concerned citizens on their days-off realistically keep pace with paid lobbyists, paid secretaries, and an unlimited supply of paper? As stipulated by committee guidelines, all members will be responsible for their own expenses and dissemination of information to their constituents. On the positive side, participation is the only option available, and it would be advantageous to keep our finger on the pulse of the lower river. On the downside, in light of MOF's lack of will and/or policy to protect this area, perhaps our limited resources might be better applied elsewhere. Will participation in LSMAC negate our efforts for protection of the lower Stikine under environment or park legislation? Not necessarily a park ... but a park if necessary.

Notwithstanding the questionable protection offered, it is interesting to note that no river in British Columbia had so far been designated under the Recreation Corridor Program.

This was not a question of some utopian nature lover being opposed to all development, or even to most development, other than the application of permanent tourniquets to one of Earth's few remaining healthy arteries. Logging is okay. Mining is okay. Flying airplanes and helicopter skiing are okay. What was not okay in my mind was the unrelenting extraction of resources with little or no regard for health of the host. Human beings are relative newcomers to this place called Earth, and Europeans are latecomers to this continent: though our intentions may be honourable, we have, by and large, yet to accept the possibility of unknown detrimental effects arising from an excessive amount of any one industrial activity or from certain combinations of industries that in themselves seem benign.

Like ourselves, the watershed ecosystem has an immune system born into it, and we shouldn't be messing with something we don't quite understand. If the ecosystem is recognized and respected as the number-one prime resource, then almost any type of industrial extraction is possible if conducted in appropriate form and scale. However, by permitting mining in parks, damming in recreation areas, and logging in scenic corridors, we seemed to have been sliding backwards while reinventing the language.

Motivated by a poll of our membership that found overwhelming support for participation, I volunteered to represent FOS in the forthcoming LSMAC process. Complementing recent studies in natural science, this administrative experience was seen as a means for furthering my education with a course in political science. Besides, I was curious to know how these things worked.

Our November '89 newsletter also gave special thanks to Director Aeron Stedman whose initiative had helped many Lower Mainland school children receive a taste of Stikine during the previous year's environment week. A student-run program saw *Save the Stikine* and *Hell and High Water* videotapes stimulate lively dialogue in conjunction with handouts of posters, postcards, and newsletters. Also, during environment week, two NFB filmstrips were introduced to local schools where favourable reviews were received from students and teachers alike. Under the heading "Tahltans of the Stikine" came two distinct works, approximately twenty minutes each, by longtime Friends of the Stikine. *From Time Immemorial* by Sylvia Albright described traditional lifestyles of the Tahltan people, introducing us to their native land and following them on their seasonal rounds. *A Sense of Place* by Gary Fiegehen examined the Tahltan people in a modern context, them trying to remain in touch with their land while seeking economic independence within an industrial society, which itself threatened their land with massive resource extraction. Wide distribution of these excellent productions into the BC school system was seen by FOS as a major focus in the year ahead.

Meanwhile, FOS directors old and new submitted reports from far and wide. With his finger on several pulses, Vice President John Christian acknowledged his acceptance of a second term as a member of the steering committee for the British Columbia Environment Network (BCEN) where he served as co-chair of the steering committee (with Nora Layard) while also being that organization's representative on the steering committee of the Canadian Environment Network (CEN). His involvement with BCEN's Sustainable Development Committee had cultivated input from environmental groups throughout the province to formulate a written brief appropriately titled *A Call To Action: Achieving Sustainable Development in BC*, from which quotes and suggestions were included in the Minister of Environment's subsequent Task Force Report, *Sustaining the Living Land*. Recognizing the ever-present need for improving communication within the environmental community, John made certain that Friends of the Stikine strengthened its links by taking out memberships in Canadian Parks and Wilderness Society (CPAWS), Western Canada Wilderness Committee (WC^2), Outdoor Recreation Council of BC (ORC), and the Northwest Conservation Act Coalition (USA).

Accompanying her partner, John, in Ottawa for the CEN's first three-day conference at Carleton University, Director Corinne Scott reported its Fair Earth conference had hosted thirty exhibitors and was attended by some 1,500 visitors, as well as media representatives. In facilitating the FOS booth with postcards, posters, and our nine-minute *Save the Stikine* call to action video, she reports many people being genuinely concerned for preservation of the Stikine while being happy to learn a river of such size remained an unpolluted and accessible wilderness experience.

Youthful energy came to us in the person of Director Karen Hodson who very capably brought back a report from the national AGM of the Canadian Parks and Wilderness Society held at Lester B. Pearson College of the Pacific in Victoria during that June of 1989.

- Keynote speaker, Mr. Thomas Berger, former Justice of the BC Supreme Court, drew on his experiences from working on the Mackenzie Valley Pipeline Inquiry to highlight the relationship native peoples have with their environment while also examining how modern technology is affecting the roles played by wilderness and wildlife in our physical and psychological universes.

- Don Ryan, speaker for the Office of the Hereditary Chiefs of the Gitk'san–Wet'suwet'en, provided some of his culture's history by describing the clan, family, and house systems, which had become key factors in court cases and negotiations. Reminding that his peoples were never conquered, nor did they ever sign over or alienate any of their lands or rights, he questioned how the government of British Columbia had assumed jurisdiction and ownership of First Nations' territories without their consent and in the absence of any treaties.

- Wilfred Jacob, land claims director of the Kootenay Tribal Council, also provided historical background on his culture and some of the devastating affects forced upon it. In looking forward, he suggested the differing gifts of all races of humankind were needed to secure a future for the planet and the organisms living on it.

- Kevin McNamee, conservation director of CPAWS, spoke to the frustrations of both Indigenous peoples and wilderness preservationists being faced with the current rate of land alienation in the interests of industrial development and economic progress. He expressed hope that our two cultures could merge their different perspectives into a common purpose.

- Larry Berg, University of Victoria, Department of Geography, immediately brought the cultural differences to light by pointing out how he, Kevin, and Tom had been speaking from prepared notes while Don and Wilfred had no notes. They spoke from their hearts—proof, he suggested, of the importance of this issue to Indigenous peoples. Larry then presented an international perspective on how Indigenous land claims and wilderness-park issues were being addressed elsewhere.

October 19, 1989
Room 200, West Block, Parliament Hill

With fingers on the pulse of wilderness issues across Canada, the Ottawa/Hull chapter of the Canadian Parks and Wilderness Society arranged a multi-media presentation about the Stikine River by Friends of the Stikine President Peter Rowlands accompanied by photographs from Gary Fiegehen and impromptu guitar music by Ian Tamblyn.

Riverfest II

Although, in my opinion, ORC's 1989 Riverfest II was more autopsy than celebration in comparison to its debut event, nine years prior, my cynicism took a back seat (temporarily) in order to enjoy *STICKEEN* **** *STIKINE*, a delightful stage production by Colin Funk's Precipice Theatre Company out of Banff–Canmore, AB, presenting a colourful look at John Muir's 1879 trip up the lower river. Bravo! An additional bright spot on that occasion was my introduction by Bob Peart to one Maggie Paquet, a lifetime writer and editor with considerable experience. Author of an acclaimed book on BC Parks, a transcriber/editor at Hansard in BC's Legislature, and an editor in the commercial book-publishing world, she had come into contact with ORC through her technical writing for the BC Federation of Mountain Clubs and through her newsletter work for CPAWS-BC. Maggie became an instant friend and the newest Friend of the Stikine. That November 1989 newsletter was her first of many fine efforts for FOS.

NEWS FLASH! In addition to the $10 million road that had been approved to run west from Highway 37 at Bob Quinn Camp and along the Iskut River to Johnny Mountain, a new push was underway for an extension to continue downstream to its confluence with Stikine. Mining activity had increased significantly in the area and the industry wanted complete access. If extended as far as the Stikine, such a road would certainly facilitate upriver shipments from Wrangell while adding a vital connection in the oft-foreseen highway link between the coast and Highway 37. Apparently, the BC Ministry of Forests was going along with the plan and was presumably following closely behind: helicopter logging by the Tahltan Development Corporation had been approved for sections of the lower Stikine. Meantime, according to then-current lip service, all development plans on the lower river required first approval (review) by the newly established Lower Stikine Management Advisory Committee. Dum-da-dum-dum.

Lower-River Research

In anticipation of my participation on that LSMAC advisory panel, I returned to the Alaska Geographic Society's handy little booklet for reliable background information, this time for specifics about the lower river. Robert A. Henning, president of the Alaska Geographic Society, opens *The Stikine River* book of 1979 in eloquent fashion:

> Any wilderness river can excite us with mystic charm, but none like Stikine. We can remember so many old ships and so many fabulous characters of another time leaving the docks in Wrangell to churn across the muddy tide flats and into the main river itself, then disappear "upriver" into a vast sea of mountains out of which came fascinating tales—of how the Tlingit Indians of today came maybe 2,000 years ago or more from their wandering trails of the Interior, through a dangerous tunnel of a great glacier to reach tidewater—of characters of many gold rushes—of vast camps of trading

> Indians—of thousands of Chinese in backbreaking toil panning the bars for ephemeral fine flour gold—of dreams of railroads, of telegraph lines, and of great mines that never happened—and a place where royalty trekked far into the wilderness plateaus to take game trophies of a wide variety. A country of history, of men, and of destiny—gateway to tidewater—gateway to the vast Interior of northwest Canada. The tides of traffic have ebbed and flooded here, and are now slack, the riverboatmen and old gold-rushers gone, a few modern miners still around with fresh hope, a slowly increasing number of tourists to a land where there are no hotels any longer, a land seeming yet momentarily ready to awake once more to new excitement and adventure. Somehow, we hope that immense power potentials do not become today's new adventure into history for the people and wild creatures of the Stikine. Let the goats have their cliffs, the Stone sheep their high ridges, the moose the willows, the salmon their tumbling torrents, and even let the roads remain a challenge and narrow and marginally maintained. Let the land sleep.

Mr. Henning and co-editor, Barbara Olds, commend their partner Marty Loken, managing editor of Alaska Northwest Publishing Company, as being the person primarily responsible for the booklet by his revisits to the river for photographs and familiarity, by his scouring of the archives, and by writing most of the final text:

> The ancestral Stikine may have been born 50 million years ago as its ancient, rocky foundation began to erode. The new river wandered here and there but settled on a 400-mile-long main channel, down-cutting and holding firm even during the past seven million years—a dramatic period in which the Stikine Plateau and Coast Mountains were steadily lifted skyward. ... The Stikine survived a series of mountain-building volcanic eruptions and repeated ages with heavy blankets of frozen ice yet in withdrawal and contributing ever more meltwater to an even greater Stikine. ... Largely ignored since its gold rush heydays, the Stikine is fast entering a new period of development which threatens to permanently alter its wilderness character. ... We have killed many of our great rivers—draining, diverting, rechanneling, damming, poisoning—but the Stikine remains, despite present threats, a living watershed, part of the last unmanipulated wilderness in British Columbia and Southeastern Alaska. We're delighted to share the river. ... Yours for a living Stikine.

A more personal look at the lower Stikine and its people was offered by Raymond M. Patterson (1898–1984) in *Trail to the Interior* (1966), which chronicles his 1948 riverboat journey on the *Hazel B* upstream to Telegraph Creek enroute to paddling down the Dease. For me, it is difficult to refrain from extracting an excessive number of passages from the delightful prose of this great storyteller; for example, the closing sentence of his Foreword does well in setting the scene: "My advice now to all who delight in northern travel is to see for themselves those rivers and the Dease Lake trail [Telegraph Road] before the improvers and the planners—all those who would destroy, recklessly and wastefully, the fair places of the Northwest—change the Stikine–Dease watershed out of all recognition."

Resting and Reflecting

11
Surging and Swerving

The initial meeting of the Lower Stikine Management Advisory Committee convened at Smithers for Halloween on October 31, 1989—ghosts and goblins to stir and to mumble, recommendations to prepare, and papers to crumple. Ably chaired by Eric Buss, MOF's regional recreation officer, our multidisciplinary panel was introduced to a pile of NOWs (notices of work)—required submissions prior to commencing any development activities in the corridor. Unclear to some of us was the relationship between the management plan and the management advisory committee. Were they mutually dependent, or could the management plan prevail on its own? Was the committee a genuine attempt to involve a wide range of concerns in the management of a priceless resource, or was it but one more bit of lip service to manage removal of the Stikine's living resources? Per Item 4.0 of the Management Plan Review: an advisory committee *should* be established for the land-use plan (which was already etched in royal cement by the Environment and Land Use Committee). It seemed the Lower Stikine Management Plan *might* (not *shall*) be beholden to recommendations of an advisory panel. Perhaps only a matter of semantics.

In Memory of Mike Jones

Words were difficult to find in sharing news about the demise of Mike Jones. His untimely death in the early winter of 1989–90 was immediately felt far beyond the familiar comforts of Tatogga Lake Resort. A big guy with a big heart, Mike was a warm-natured multi-dimensional man whose passing left a big hole in many lives. A close neighbour of his, Wade Davis, offered appropriate testimony in his book *Shadows in the Sun* (1992): "Mike, besides being as good a bush pilot, riverboatman, and hunting guide as the country had ever seen, knew how to spin stories that, as the old Indians said, could make the stars shift in the sky. …

Sometime just before Christmas, Mike disappeared on his trapline at Bowser Lake. The Mounties found his dog, Chinook, 35 pounds underweight and judging from the few pelts found in the line cabin, they figure Mike went through the ice not long after he reached his trapline. They didn't find his body until spring. The entire valley went numb with grief and disbelief. Mike knew the bush as well as anyone. He understood the whispered messages of the wind."

Our second gathering of LSMAC on April 12, 1990, also in Smithers, found few items in need of attention for the corridor proper: permission for use of an already installed BC Hydro sensor and a notice of *interest* (not work) by Stikine Copper for areas near the Scud and Anuk Rivers. With FOS and Tahltan Tribal Council (TTC) as minority dissenters, LSMAC gave approval to the application by Great Glacier Salmon (GGS; formerly GGFishing) to construct a number of duplex units on-site to provide family accommodations for staff … with a proviso banning their future use as a tourist facility. As identified in our letter to Chairman Eric Buss (January 31, 1990), FOS's objection to the GGS expansion was based on the fact that it was contrary to the wishes of the local people, and because it was thought premature to be approving any development plans before some refinement in the management plan's direction: this otherwise benign application could be the thin edge of the wedge opening the door to a greater number of applications for significantly larger projects before LSMAC had defined itself with consensus on the broader issues.

In the meantime, plans were going ahead for the Tahltan Nation Development Corporation's proposal for selective helicopter logging in the corridor's Timber Management Unit #3—Darsmith Creek to the Alaska border—utilizing aerial methods only *at this time*. Also discussed at this meeting was the current situation regarding "snag crews," which for decades had been responsible for maintaining a navigable channel on the dynamic lower river. While river users were concerned that recent budgetary realignment had decreased the snag crews' effectiveness, Ministry of Environment agents were increasingly concerned about potential damage to fish habitat and natural drainage patterns caused by the requisite blasting and cutting. While considering the channel-clearing contract to be subject to LSMAC review as were all "works" in the corridor, the matter was deferred, pending more interagency information sharing.

As further reported in the summer 1990 newsletter from FOS, agreement in principle had been reached between BC and Alaska for a resource highway connecting the Iskut Valley gold fields with tidewater at Wrangell. This proposed road would come up Bradfield Canal and then follow the Craig River Valley downstream to Johnny Mountain, Snippaker, et al. Another road, already approved by BC, would connect the Bronson Creek airstrip (Johnny Mountain) to Highway 37 at Bob Quinn Lake by following the lower Iskut River. A spur from this road would connect with the other impending gold property at Eskay Creek.

Impatient to open their recently approved Snippaker Mine and apparently unable to afford their share of road construction costs, Cominco had recently purchased a twelve-tonne capacity hovercraft to service its property. According to project manager, Merlin Royea, Cominco's hovercraft would haul ore down the Iskut and lower Stikine to the Alaska coast where it would be offloaded into barges. Those plans could change if a road were built, he said, but even if it were to happen, the hovercraft would be used until the road opens. Described as an air-cushion vehicle leaving no wake, it was considered *less detrimental to birds, mammals, and recreating humans than barges and boats*. A notice of work had not been submitted to LSMAC, whose corridor mandate included several miles of the lower Iskut as well as the Stikine.

From December 9 to 10, 1989, I was at Econiche House, a small conference centre in the snow-covered hills of the Gatineau Region, just north of Ottawa. CPAWS and the CEN had coordinated a meeting of more than fifty individuals from various environmental groups to share concerns while working toward a national wilderness strategy. Following a mostly frustrating day of presentations, waiting for direction to emerge from the spinning of many high-energy wheels, the evening cross-country checkup was dismal. Les Parcs du Québec, large green areas on the map, were not really parks, only large forest preserves where significant physical force had been applied to Indigenous families protesting deforestation of their lands. Nearby, the Cree people of Mistassini were reported to be in horror while watching thousands of caribou drowning behind the dams of La Grande.

Over in Ontario, roadblocks and rhetoric were continuing for the remaining six hundred hectares of red and white pines in Temagami while some hope remained for a new wilderness conservation commitment to prevent logging in Algonquin Provincial Park. The situation was equally desperate in Manitoba, Saskatchewan, and Alberta, where multinational corporations with Canadian-sounding names had secured rights to clearcut massive tracts of boreal forest for pulp production. Derek Thompson of BC Parks reported that, with a very sudden and very recent spate of legislative protection for many of its parks, the province now boasted one of the largest and best protected provincial park systems. Nevertheless, behind Brazil, British Columbia was then recognized as the second largest defoliator in the world: decrying forestry practices elsewhere, we Canadians were unabashedly destroying our forests and subjugating Indigenous peoples.

It seemed, with help from free trade and the Meech Lake Accord, huge sums of money were being consumed in the name of "sustainable development" by wealthy parties who didn't know a watershed from a woodshed while local conservation groups were scrabbling and begging for postage money.

> Trade has always been free
> It's the reality of transportation
> And the fantasies of enterprise
> That cost us so dear
> ?
> Who takes care of the land while
> Government takes care of business.

On the following day, our gathering identified itself as the Wilderness Caucus and got down to business. While the Valhalla Society was putting forth a proposal for a nationwide extension of its BC endangered spaces mapping process, Grant Copeland presented the motivational videotape *Vanishing Wildlands* (previously previewed by FOS at its AGM). A subcommittee of three—Laurie Montour, Assembly of First Nations; Tom Munson, Yukon Conservation Society; Peter Rowlands, Friends of the Stikine—collaborated on a resolution recognizing the role of Indigenous peoples and the need for resolution of land-claim issues toward sound resource management in Canada.

Monte Hummel of the World Wildlife Fund shared a copy of a letter from then federal Environment Minister Lucien Bouchard who had committed himself to completion of the Canadian national parks system, notwithstanding the fact he had stood *with* his government members in the House of Commons to *defeat* Skeena MP Jim Fulton's motion, which sought commitment for protecting twelve per cent of the national land base by the year 2000. Indigenous rights, militarization of the north, and the wholesale desecration of our boreal forests were the major issues recognized by the caucus. CPAWS, WWF, and *Borealis* magazine agreed to collect and disseminate information, maybe with a *computer* network. We would meet again.

Meanwhile, back in Beautiful BC, the provincial Outdoor Recreation Council (ORC) under direction of Bob Peart, its new executive director, was addressing key changes recently made to the wilderness policy issued by the Ministry of Forests. Generally, these changes by MOF had opened up designated wilderness areas for activities normally considered incompatible with wilderness: contrary to the initial draft, new grazing permits[18] were now *allowed* in wilderness areas; contrary to the initial draft that had *prohibited* motorized vehicles, the newer version allowed for *restricted use* on land, sea, and air under vague controls; the minimum size for designated wilderness areas had been reduced from five thousand hectares to one thousand hectares. Calling for creation of an advisory committee to guide it and for inclusion of the term *roadless* in its wilderness policy, ORC brought forward a reminder of the working definition adopted by the original 1986 Wilderness Advisory Committee: *Wilderness is an expanse of land preferably greater than 5,000 hectares retaining its natural character, affected mainly by the forces of nature with the imprint of modern man substantially unnoticeable.*

At a meeting in Terrace to discuss parks in northern BC (March 6, 1990), FOS made a presentation to Parks Minister Ivan Messmer, congratulating him and his department for the best park system in Canada while encouraging him to make it even better. Given the fact that many of our watershed ecosystems had been totally dedicated to development, so too should some be left in their natural state. On this premise, we recommended four elements for change in the Stikine: 1) Mt. Edziza park expanded northeasterly to include the Grand Canyon of the Stikine, and Spatsizi park expanded southwesterly to include headwaters of Spatsizi River; 2) Creation of a "headwaters" national park reserve to also include upper reaches of Nass and Skeena Rivers; 3) Canadian heritage river designation for the entire Stikine mainstem above the international boundary; 4) Stikine Watershed Biosphere Reserve under auspices of the IUCN.

Predictably, our presentation included detailed rationale that made perfect sense to us. Included were wide-ranging personal essays on both the upper and lower rivers, which rambled on from bits of history through current threats and resolutions into holistic visions for the future, the essence of which can still be found in bits and pieces scattered throughout this account.

Back in Vancouver, May Murray and I participated in a series of meetings for ORC's Rivers and Shorelines Committee, which used resolutions from the 1989 Riverfest II to begin working toward a coastal water strategy and a provincial wild-rivers strategy while also examining the potentials of recreation corridors and heritage river systems. With another BC rejection of CHRS announced by Parks Minister Messmer, ORC's Rivers and Shorelines Committee decided to develop its own wild rivers strategy for presentation to the government. For many of us, it was difficult to understand why or how a provincial government ostensibly interested in honouring its rivers would not participate in CHRS. At the time, any commitment to watershed integrity, however small, would have made the multiple-use vision for our remaining ecosystems seem more palatable. In any case, given the size of the industrial lobby and the speed of its northern progress, haste was certainly in order. For its initial contribution to ORC's wild rivers strategy, FOS selected seven of the wildest ones that required immediate attention: Tatshenshini, Stikine, Spatsizi, Dease, Khutzeymateen, Babine, and Stein.

Other then-current items of interest: an electric energy forum by BC Hydro and MOE in Prince George; an essay on Stikine–LeConte Wilderness Area by the ranger for Wrangell District; notice of a major moose–wolf research project in Spatsizi; a Project Wild conference in Hawaii co-sponsored by WC^2; a Voices of the Earth global conference on Indigenous peoples and the environment sponsored by the Lytton Indian Band and Mt. Currie Indian Band; and a sixth Stein Valley Voices for the Wilderness Festival with Gordon Lightfoot again as first confirmed participant.

Filled with such items, Maggie's first-of-kind newsletter was printed magazine-style in simple black and white format on lightweight paper. As well as the latest news, this first-of-its-kind Summer 1990 issue also carried a new title—*THE CURRENT*—with a nameplate in the form of a graphic wave pattern containing the "friends of the Stikine" identifier and its iconic water droplet. The front cover featured a fine pencil rendering of the RiverSong by Bill Wheeler, an accomplished artist and illustrator, who had visited Stikine the previous year with a canoe party from York University led by Larry Hodgins (a Wanapitei brother with Bruce). In a letter summarizing their thoroughly enjoyable and somewhat exciting experience, Bill Wheeler added, "I agree wholeheartedly with the aims of your group to keep the Stikine unspoiled. It has so much to offer that it would be criminal to destroy any of it."

12
Bouncing and Barging

Modern Man Manages Lower Stikine. This headline-worthy quote erupted onboard a heavily loaded river boat returning to the dock at Telegraph Creek with an abundance of blue smoke billowing from its engine compartment. It was one of those days, and the quote could have been the overall title for our ill-fated field-trip adventure with the Lower Stikine Advisory Committee (now LSAC, after officially dropping "Management" from its title). The disappointment experienced by the half-dozen committee members on board was nothing compared to that of owner–pilot, Richard Royea, who had trailered his boat in from Dease Lake, or that of committee chairman and trip organizer, Eric Buss who watched another brilliant piece of crisis management go up in smoke. His itinerary to get thirteen of us on the river for two days had run into nothing but headwinds. Last-minute cancellations had left him without a large riverboat and without an Otter support aircraft. Richard's replacement vessel was barely adequate, and arrangements had been made with Ron Janzen of TelAir to charter his Cessna 185 out of nearby Sawmill Lake with excess people and supplies: he would rendezvous with the committee somewhere on the river to convene for the night at Great Glacier as originally planned.

The actual river trip lasted about five minutes. It could have been worse. The failed water pump and fried piston might not have appeared until we were well downriver in a location inaccessible to Ron's 185. Chairperson Eric nonchalantly implemented Plan D: our gear was offloaded (again) and reloaded into the ministry's green Suburban vans for transport down the Glenora Road where we camped for the night at nearby Winter Creek (aka Four Mile), which offered a fine haven on the banks of the Stikine. Meantime, sitting idle amid all this activity was the ideal replacement boat, but its owner, Dan Pakula, a local resident and committee member, was in downtown Smithers with his wife, Diane, for birth of their fifth child.

Next day's inspection phase of our mission was successfully accomplished by another brilliant (and expensive) decision by chairperson Eric and Lauren Robertson of MOF Dease Lake: a locally based helicopter and TelAir's 185 were chartered to provide everyone an airborne look at the corridor to examine six potential recreation sites, four proposed and/or existing developments, and five notices of work.

Potential Recreation Sites: Great Glacier (River) was, by then, already an organized campsite in development by MOF with firepits, tables, biffies, and a trail to the lake for viewing the glacier—a natural and highly appreciated facility. Choquette Hot Springs, situated nearby across the Stikine, was recommended to remain low key and without development other than an access trail. Scud and Anuk snag crew cabins were considered fine just the way they were. While level expanses such as Hudson Bay Flats and Jacksons were seen to be suitable for recreation site designation, a strong recommendation was made for wild and naturally good campsites such as Yehiniko to remain exactly that—unmarked and unidentified for river travellers who prefer to "discover" their own remote and unprepared campsites.

Proposed/Existing Developments: Viewed from the air, Great Glacier Salmon and the Tahltan Nation fish processing facilities looked to be coarse and chaotic incursions in the Stikine River shoreline (not looking so bad from the river in shitty weather). There was no obvious sign of Bob Gould's recently approved condominium-duplex-subdivision at GGS. Meanwhile, an application by Bill and Ruth Sampson for a fish-landing site and storage facility nearby was under review. Under the heading of "NEWSPEAK," the old clearcut at the Iskut confluence, long ago sanctioned by MOF, was now being referred to as Halvorson's Log Sort—slowly but surely disappearing from the viewshed and the watershed through bank erosion. No surprise there.

Notices of Work: There were literally scores and scores of geological/mining worksites in the lower watershed, mostly located in the Golden Triangle southwest of Mt. Edziza between the two main rivers. Although two large mines were in operation over on the Iskut, there was little activity seen in our corridor other than cut lines and small trenches for sampling. In need of fuel, our returning helicopter made a pit stop a slight distance up the Porcupine Valley where we found a fixed-wing airstrip, several fuel caches, a satellite dish, and a cluster of tent-frame igloos. Some surprise here! Previously unknown to any of us, including our mining industry representatives, Porcupine Strip was soon "discovered" to be a long-term base camp for a mineral exploration group (Pezim et al.). Although beyond corridor boundaries and outside our mandate, the presence of Camp Porcupine a few miles off the lower Stikine in possible salmon spawning habitat underlined the inadequacies of our advisory process and its established boundaries. Drawn solely to satisfy visual retention parameters of a logging plan, our boundaries reflected only river-level perspectives and excluded many other values.

The recently retitled Lower Stikine Advisory Committee retreated from the river to conduct its annual review at the BC Ministry of Forests office of the Cassiar District in Dease Lake on September 12 to 13, 1990. The first order of business was a brief appearance by Tahltan Tribal Council President Pat Edzerza who read a prepared statement declaring the Lower Stikine Advisory Committee to be without jurisdiction over the Tahltan people and without authority in their traditional territory. On behalf of the TTC, President Edzerza said he would conduct any future discussion only with Eric Buss[21] or directly with government ministries. Perhaps, as revealed in a later order of business, MOF's recent decision to delay TNDC's (Tahltan Nation Development Corporation) application to helicopter log in TMU#3 may have been a factor in the TTC's then-current position.

While recognizing the possibility of its first annual field trip also being its last, LSAC remained on task. Yes, the outstanding Tahltan land claim was easily seen as the foremost obstacle to effective management of the lower Stikine (by us). Along with the need for more attention to corridor boundaries, we expressed concern for the possibility of "paper fences" applying a false sense of protection around recreation areas in a sea of industrial development: although the lower Stikine is forever recognized as a working river, it would be a serious mistake to manage all the wildness out of it.

Committee member and lower-river resident Doug Blanshard had been left high and dry by our aborted river trip, but he managed to join us in Dease by catching an airplane ride from his cabin on the lower river. Doug was the only truly local input to our field trip. "Yes," he said, "Cominco's hovercraft is already operating ... a large, noisy machine, a difficult-to-miss detraction on the lower Iskut and Stikine ... about two daily return trips between Bronson Creek and Wrangell. As advertised, it produces no boat-like wake ... but pressure-induced swells are noticeably affecting shorelines and threatening to undermine existing structures as well as creating grave concern for salmon spawning habitat." The hovercraft had obviously slipped through the wide-open cracks in our rustic bureaucracy: having apparently obtained all necessary operating permits, Cominco had seen no need to notify our fledgling committee of its use ... and they would probably never-ever seek its permission. *Having been declared environmentally benign long before its arrival, the hovercraft had already become a fact of life on the lower Iskut–Stikine.*

Editorial comment from *THE CURRENT* newsletter for Winter 1990/91: "You don't have to be a rocket scientist or president of the tribal council to know our humble LSAC committee had no authority. Almost as an afterthought, the panel was created as an appendix to an industry-induced management plan which circumnavigated its own public review process on the way to cabinet approval. Our 'monitor and advise' mandate is as vague as our relationship is to the management plan itself. So far, we had 'managed' to impose another layer of paperwork on local people wanting to upgrade their properties on the river while hovercraft and mining activities continue to accelerate within the terms of the management plan."

Business was booming. The degree of enthusiasm in the area could easily be measured by the amount of aviation activity. Central Mountain Airlines (CMA) out of Smithers and Trans-Provincial (TPA) out of Terrace each operated scheduled return flights at least three days per week to Iskut, Telegraph Creek, and Dease Lake as well as on-demand charters to numerous smaller resource strips. In addition, TPA operated two Bristol Freighters between Bronson Creek and Wrangell—fuel and supplies in, ore and concentrates out—while also operating a Convair passenger charter each week between Vancouver and Bronson. CMA operated a daily DC-3 between Smithers and Bronson with men and supplies while a steady stream of ore samples flew south from Smithers via Canadian Airlines. A minimum of six full-time helicopters were based in the area during the summer months to support mining activities and probably twice as many were there during peak periods.

"What is the point of public input?" asked Monty Bassett, in writing to the *Interior News* of Smithers: "The Emperor has no clothes, and the BC Forest Service has no credibility, at least none when it comes to public advisory input." Recounting his recent participation in a demanding local process which had claimed public input to be vital, Monty reported the window-dressing fears of local people to have been blatantly realized—their recommendations totally disregarded as part of a province-wide phenomena. "Our forests are managed for the industry's interests, not for the public's, not for the logger's, not for regrowth possibilities, not for wildlife habitat, and not for tourism. Our forests are being harvested for maximum profit. According to MacMillan Bloedel's annual reports, their profits rose in a two-year timeframe from $43 million to $330 million! Yet jobs and wages have not increased at all. Unfortunately, our forestry practices aren't even labour intensive: they're bank intensive and corporate intensive. They're loan intensive and debt intensive, but not human labour intensive."

> ***Fancy cutting down all those beautiful trees to make pulp for those bloody newspapers and calling it civilization*** —Winston Churchill, touring Canada in 1929.

Reproduced here, thirty years later and in a mercifully edited state, my mini rant captures the frustration of a newsletter writer attempting to wrap his head around that reality:

> Traditional lands of the Tahltan People are swarming with hunters, fishermen, photographers, canoeists, kayakers, rafters, tourists, prospectors, geologists, miners, loggers, engineers, and environmentalists. The first clearcuts have long been made and numerous mines are in steady operation. The Lower Stikine Management Advisory Committee has yet to receive an application for consideration of Cominco's 80-foot hovercraft which is **already** operating within LS(M)AC's mandated area of responsibility. Upstream of Canada's Grandest Canyon, the BC Parks ministry, in the interests of public safety, has begun installing signage along one of the world's wildest and most scenic five-day floats. Ugh! Wilderness advocates are caught between a rock and a hard place—not wishing to impose their vision on the land-based local economy and unable to influence the status quo of the resource-extraction oriented provincial government. Subject only to their own in-house environmental assessments, resource extractive industries continue almost unfettered.

On the positive side, FOS received a heartwarming letter from an anonymous geologist working in the Stikine who reported his camp to be "minimum impact" with all recyclable items returning on CMA backflights to Smithers and being donated to a non-profit organization. After reading issues of THE CURRENT and *Borealis,* he avowed better understanding of threats created by mining and wished to see better balance between industry and environment. He expressed disappointment in the BC government for bending so completely to the wishes of large mining and logging companies and felt the Stikine Valley had far greater value than just for mining purposes. Thank you, Mr. Anonymous; it was refreshing to learn there was another way to approach the business of resource extraction, and to know that respect for the land can and does exist within industry.

By the Spring of 1991, I was, for the most part, in peace. Gulf Canada's Mount Klappan coal project at the headwaters was still an open question remaining subject to markets, financing, and approval. In the Grand Canyon, BC Hydro's buildings and facilities were being removed or modified by BC Parks in an effort to naturalize that portion of its recreation area. While that benevolent utility continued to simmer its five-dam megaproject on a back burner, two small-scale hydroelectric projects had been approved to replace diesel generators for local needs at Dease Lake and Iskut. Downstream, the jury was still out on the impact of Cominco's hovercraft. In fact, a study had never been actioned. Cominco seemed happy enough with its hovercraft to be uninterested in paying for a share of the proposed mining road from Highway 37 down the Iskut valley to Bronson Creek. At nearby Johnny Mountain, Skyline's operation was in limbo because of uncertainty about available reserves. The new kid on the block, Eskay Mines, was poised to benefit most from ground transportation because of ore reported to be highly acidic.

Ta-da! An announcement by Dave Parker, BC's new minister of Lands and Parks, soon gave the go-ahead for a *publicly* subsidized road from Highway 37 down the Iskut as far as Volcano Creek, thus catering to an access road for the Eskay Mine, situated slightly over the divide in the Unuk watershed. Given this mine's high elevation and reportedly high acid content, road access seemed necessary. Given distance and terrain factors, the idea of road access up the Unuk from tidewater was pretty far-fetched: assuming the Unuk River to be typical high-quality salmon habitat, an upriver road there might be as big a threat as a downstream washing unit at the mine site. Although the proposal for a similar road from tidewater to Bronson via the Craig River had been scrapped for the moment, concern remained for the possibility of Parker's new public road being pushed farther downstream to Bronson. From there, a major fear was for another cabinet shuffle to see the road extended down to Johnson's Landing near the Stikine. From there (horror of horrors!), another realignment of economic realities might see that same road extended upstream on the mainstem Stikine for access to the mineral-rich Scud and Porcupine areas. So far, the eternal dread of "scenic" highways along the lower Stikine were being held in check by the Stikine–LeConte Wilderness area on the coast. God Bless America.

The challenging terrain of the Stikine watershed is more than simply an added expense for industry—it's an integral part of a living process helping to sustain humankind. The stock market may be important, but it's not the only market. A healthy pension plan is of little use without fresh water to drink. And no, we can't survive without the birds and the bees ... and the mosquitos.

> ## Perseverance
>
> Among directors' reports in *THE CURRENT* of May 1991, Monty Bassett reported the first (known) unassisted descent of the Grand Canyon of the Stikine. Paddler Doug Ammons reported, "It was both a kayaker's sweetest dream and his worst nightmare." Regarded as the Everest of Whitewater, Stikine's Grand Canyon had long held the fascination of team leader Rob Lesser who credits an overflight in 1977 with saving him from excruciating embarrassment if not certain death. Attempting the canyon in 1981, he and his paddling partners had to bail out halfway down the canyon when their filming budget ran out. Returning in 1985 with a better budget and a helicopter, Lesser, along with Bob McDougall and Lars Holbeck managed to technically paddle the entire canyon while helicoptering out to base camp each night. (See *Hell and High Water,* 1985.) At the time, Lesser believed an unassisted run might be possible and was soon planning an attempt. In 1989, with McDougall and Ammons, a freak accident early in the canyon caused loss of a kayak and a long day-and-a-half hike out. In September 1990, after surviving a failed automobile and a ripped dry suit, Lesser, Ammons, and Tom Schibig were successful, dropping in at the Highway #37 bridge and pulling out at Telegraph Creek, totally under their own power and unassisted. Schibig says they took no chances, but sometimes they had no choice but to run certain sections: "All in all, we had to portage six places, and of those, we possibly could have run four." No helicopter backup this time. Once again, Lesser marvelled at the canyon's dynamic whereby huge slabs detached from the walls and the river continued to change in nature: "Some rapids we were worried about in 1985 had disappeared completely, but others materialized where there were none before."

Down south in Greater Vancouver, Director Corinne Scott reported excellent feedback from many people attending two four-day events where she hosted our FOS information kiosk. Total attendance for the RV Show and Great Outdoors Show was estimated to be in the neighbourhood of fifty thousand, many of whom stopped by to be entertained by video presentations and slide shows courtesy of Tony Shaw and Gary Fiegehen. Along with books, maps, visitor guides, posters, postcards, and buttons, the latest hot items for sale over the counter were Great River T-shirts featuring Gary's now-somewhat-famous image of Sergief Island. Corinne said she had been surprised to hear from so many people who had lived and/or visited Stikine as well as from numerous others who claimed to have ready-made plans to visit. "Thanks to the organizers of both shows for making affordable booth space available to environment groups," added Corinne. "Without that help, we certainly wouldn't have been able to participate. See you next year!"

Director Maggie Paquet attended BC Hydro's Fourth Annual Energy Forum in Victoria on April 16 to 18 and had reconnected with favourite university professors while participating as facilitator in a series of talks and discussions with primary focus on energy education. A fairly broad agenda visited subjects such as conservation ethics, environmental awareness, and bringing science to life. In sessions with Senator Michael Kirby relating to public attitudes, Chris Boatman, VP Corporate and Environmental Affairs for BC Hydro, declared the old days done and gone: Hydro could no longer get away with only telling the public what it was doing after it was underway; BC Hydro said it had realized the need to communicate and agreed to it being a two-way dialogue about real problems and possible options. "It's an attitude that shows great promise. I'm sure most of us in FOS would encourage BC Hydro to build upon these inspiring words," said Maggie, while thanking them for the financial assistance which had enabled her to attend and to participate on behalf of Friends of the Stikine.

Director Karen Hodson did a wonderful job recording the proceedings of our 1990 AGM, held January 26, 1991, in ORC's conference room, in which she summarized reports and activities of all directors, including Don McClure at the Heritage Forests Society; Corinne Scott at the CEN AGM, the Okanagan Film Festival, and the Environmental Youth Conference, while also setting up a Stikine Watershed display in the main window at Mountain Equipment Co-op; David Frost at ORC's Rivers and Shorelines Committee, seeking to establish a provincial rivers strategy; and John Christian, seeking financial assistance from the Shell Environmental Fund. May Murray showed us to have 145 members (forty per cent from outside Canada) and a mailing list of 330 which included thirty-four other environment groups and twenty-two government agencies. Joining those above, newly elected to the board were Dan Pakula, Kim Pechet, and Jon Shemerdiak.

Featured guest Mark Angelo,[13] chair of ORC's Rivers and Shorelines Committee, said the *U.S. Wild and Scenic Rivers Act* would be used as a guideline in formulating a strategy to be presented to the provincial cabinet for recognition of BC's scenic, wild, and recreation rivers. Given BC's aversion to CHRS at the time, he suggested the need for a Wild and Scenic Rivers Coalition for BC and maybe for Canada. Despite contrary remarks from the local MLA, Mark considered the Site C project on the Peace River to be "permanently" on the shelf while BC Hydro turned its attention to smaller independent power projects (IPPs). The times do change, don't they?

NEWS FLASH!!

By time of newsletter printing, we had been awarded $4,000 from the Shell Environmental Fund to facilitate travel arrangements for our busy summer of meetings planned in Stikine Country. Our sincere *thank you* to Shell.

The subsequent FOS newsletter, *THE CURRENT* of September 1991, opened with a President's Report about the spring LSAC meeting at Hudson Bay Lodge in Smithers (May 22, 1991) where foremost on the agenda was discussion of an increasingly complex situation arising from new rules under the General Agreement on Tariffs and Trade (GATT) which now allowed salmon, after being first landed in Canada, to be then sold in Wrangell, USA. Application for a fish-landing station by Stephan Jacob (a founding partner who had withdrawn from GGS) had excited controversy involving Bob Gould (GGS), and the Tahltan Fish Company against Jacob and other independents such as Bill Sampson. Threats of legal action were exacerbating the situation. Bob Gould attended this LSAC meeting and presented his version of the story which helped to clarify events, if not their motivations.

Short story: Gould and company were disturbed by the prospect of Jacob and Sampson et al. being allowed to construct residential fish-landing stations that were denied to individuals in the GGS group—no *residences* had been approved there, only duplex-style accommodation for employees of that one fish-landing site. Obviously, there existed differing views on competition and fair practice in the eyes of some free enterprisers. While watching modern man managing resources of the lower Stikine, FOS and other committee members reiterated the call for clear and simple corridor definition and direction before being asked to adjudicate land-use applications of unknown extent. Otherwise, there existed the possibility of unrestrained development activity defining, if not dismembering, the committee itself—as in the tail severely wagging the dog. To create breathing space, LSAC decided to freeze all land applications on the lower Stikine for one year, allowing time for resolution of some on-river differences and giving LSAC time for getting its act together before being swamped by its own mandate. Unfortunately, we were already a bit late in the game.

Cominco's giant AP1-88 hovercraft was already operating two or three return trips per day between Bronson Creek and Wrangell and, in the process, proving previous reports of increased bank erosion to be well-founded: it seemed, especially at high water levels, the machine's high-pressure swell was having noticeable effect. As one might expect, the *Stikine/Iskut River Hovercraft Operation: Environmental Summary* prepared for Cominco Metals made no reference whatsoever to the threat of wake/swell in a narrow channel while concentrating largely on noise levels. Having been declared environmentally benign prior to its arrival, this air-cushion vehicle (ACV) had become a fact of life in the lower Stikine. While our LSAC committee had been forced to suppress or curtail the activities of several long-time residents, a far-distant corporation was being allowed to operate full speed ahead on the lower Iskut–Stikine.

Also introduced at this September LSAC meeting was a document entitled *Lower Stikine River Recreation Study* prepared by Margaret Churchill—an excellent first-ever attempt to identify potential recreation sites and to categorize shoreline carrying capacity along the lower Stikine. An invaluable tool for the task of LSAC, this biophysical study was also a step toward preparation of an interpretive brochure envisioned by MOF. In it, units of river and uplands sharing similar characteristics were identified in five corridor segments—Glenora, Chutine, Scud, Glacier, Iskut—all described in broad terms of climate, soils, vegetation, morphology, and terrain. Within each segment, specific sites were evaluated according to carrying capacity and recreation potential. After discussion on four possible development options, our committee opted for a compromise arrangement that provided some map reserves (including sites and trails) while reiterating caution about managing away all the wildness. More comprehensive than its title suggests, this study was a giant step forward. Ms. Churchill was gratefully thanked for her thorough research and for timely completion of a comprehensive guide to the lower Stikine River.

Friends of the Stikine concluded activities for 1991 at its AGM on November 23rd, courtesy once again of Blair Shakell and his studio space where we were entertained by a two-projector musical slide show featuring images from Gary's recently released book, *Stikine: The Great River*. As well as reports from directors, the meeting resolved to initiate a hunt for professional assistance with policy and bookkeeping, motivated by a letter from Revenue Canada calling for an improvement in our performance with regard to our charitable society status. An audit of our organization covered the Expo 86 period when, like many British Columbians, we (one of us) tended toward excess (donating fuel costs for charter flights, etc.). Although this AGM produced little change in the list of directors, it did see me step down from the president's chair (an action not necessarily related to the previous item) in order to address overdue personal issues while May Murray and Maggie Paquet remained co-chairs of an administration promising to be better than ever. The fact that May and Maggie had been holding the show together for some time promised a smooth transition to 1992—as suggested by a river symposium photograph of Maggie, me, and May.

HISTORY 303

Although furs were the primary concern of white men first approaching the mouth of the Stikine, it was gold—irresistible and elusive gold—that brought the first steamboat to the river following unknown centuries of human-powered canoes used by the Tlingit people. Gold was discovered in the vicinity of Great Glacier in 1861 by Alexander "Buck" Choquette who is remembered on the map by Buck's Bar and Choquette River (and its hot springs pond). Word of the discovery quickly spread to the Fraser River diggings and reached the ears of a Prussian with the unlikely name of William Moore who was steamboating on the lower Fraser with a fast sternwheeler named the *Flying Dutchman*.

Late in the spring of 1862, Moore loaded a barge and his steamer with 125 passengers and assorted cargo and headed north toward Alaska, a Russian-owned territory being leased to the Hudson's Bay Company, which had since abandoned its Fort Stikine trading post (which became Fort Wrangel on its way to becoming Wrangell). By means unknown—perhaps with a hired Tlingit guide—Moore pushed the *Flying Dutchman* up to the head of navigation (however far he could go) for several trips over seventy-two days, netting him $14,000, which was more than any of the miners made. Two years later, that "Strike of '61" was recorded as a fizzle, each miner panning only $3 to $10 per day. The Dutchman flew back south to the Fraser and there were no steamboats chugging on the Stikine for the next four years.

Then along came the telegraph line: Western Union–Collins Overland, on its way to Bering Strait, had the sternwheeler *Mumford* built at Victoria to facilitate construction logistics on the Skeena and Stikine rivers. During the summer of 1866, large quantities of wire and equipment were deposited at the head of navigation on the Stikine, at the mouth of a small stream that became known as Telegraph Creek. The successful laying of a trans-Atlantic cable in the following year saw this overland telegraph line abandoned and much of its equipment re-shipped downriver as scrap. By then the Stars and Stripes had been raised on the coastline of the far northwest.

The lower river was quiet again for another seven years until a pair of hardy prospectors, Henry Thibert and Angus McCullough (or McCulloch), travelled by an overland route to discover gold at Dease Lake in 1874, inciting the biggest rush since the Cariboo excitement of the early 1860s. When Captain Moore returned to the lower Stikine with three successive boats, he found some competition: Captain John Irving, a famous name in BC marine history, was there with the *Glenora*, one among several operators now shuttling between Fort Wrangel and the relatively new community of Telegraph Creek, both of which had become major supply and transportation hubs. John C. Callbreath of Telegraph Creek established a trading business there that ran boats on the river while also operating Alaska's first salmon hatchery.

Then came the Klondike Gold Rush. In the spring of 1898, the streets of Fort Wrangel were busy with eager gold-seekers who provided good business for sternwheelers that deposited them and their gear on the ice at the river's mouth—a head start on the all-Canadian route north from Glenora where plans for a railway had already seen several warehouses built. Not a pretty picture: thousands of men in a wretched little tent city gambling away their grubstakes until the snow had become too soft to walk on. When the ice went out, there were sufficient steamboats on the river for Canada's Department of Marine and Fisheries to establish a lengthy list of navigation rules. It was a mad, pell-mell business (while it lasted—for all of two months), jostling for position, winching up narrow canyons, and clearing the riverbanks of trees for fuel. When the winds of reality blew all the hype out of the all-Canadian route, most of the steamers dispersed to work somewhere between the Yukon River and the Fraser … or to the boneyard.

The Hudson's Bay Company continued providing several trips per summer to Telegraph Creek for big-game hunters and big-dream miners; otherwise, the lower Stikine returned to a quieter and more relaxed state. In 1902, Fort Wrangel officially changed its name to Wrangell while becoming more of a sawmilling centre and fishing port than international transportation hub. The steamboat era on the Stikine ended in August 1916 when *Port Simpson*, a large vessel of the Hudson's Bay Company, completed her last trip to Telegraph Creek before being withdrawn from service for financial reasons.

In the same year the *Port Simpson* pulled out, a new figure appeared on the Stikine, a person who was to dominate river traffic for many years to come. Having made his reputation on the upper Yukon in the gold rush days, Captain S. C. "Syd" Barrington and his new *Hazel B. No. 2* barged down from Anchorage to be joined by his long-time partner, Captain Charles Binkley, and by his brothers Hill and Harry Barrington. For thirty-five years or so, they operated six motor vessels on the Stikine—all of them named *Hazel B.* in honour of Syd's wife—supplying ranches, trading posts, and small-scale mining outfits, as well as casual passenger traffic such as trappers, prospectors, tourists, and big-game hunters. Then along came the airplane and a road to Telegraph Creek. Commercial boat traffic on the lower river dwindled to almost nothing.

The first steamer on the Stikine, the *Mumford* of 1866, had been a wood-fired sternwheeler, 110 feet long with a beam of nineteen feet. For historical comparison, the third and last *Hazel B. No. 2* iteration of the 1940s was one hundred feet long with a beam of twenty-five feet and driven by retractable twin screws powered by two 135-horsepower diesel engines. With refrigeration and electrical systems, she could seat thirty people in her dining room while drawing but twenty inches of water when fully loaded.

In little more than the length of one human generation, boat traffic on the lower Stikine had progressed from poles and paddles through wood and steam to gasoline and diesel.[20]

LOWER WATERSHED

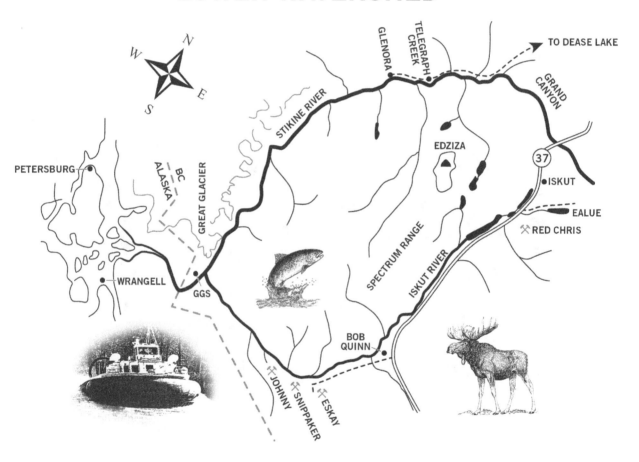

13
Swishing and Swirling

Perhaps practical minds prefer to avoid thinking about wisdom and wilderness because neither is subject to human management. They happen by themselves, according to natural processes that are not understood. No educational system knows how to create wisdom, and no science can make wilderness. We know how to damage and destroy both of them, however, and we have devoted much of our energy to that in recent centuries. Before we reach the point where both wisdom and wilderness cease to exist, we should think about what they are, how they relate to one another, and what the world would be without them.

These words by Joseph Meeker from *Comedy of Survival* (1980) are apt entrance into Gary Fiegehen's book *Stikine: The Great River* (1991), which presents photographic overview of a significant ecosystem in its near-natural state. The book's textual addenda offer stories from the oral tradition alongside written history while including occasional reference to possible future threats. Using an assortment of high-quality photographs over a broad spectrum, Gary gives us a good look at what was there then. It is a portrait of no specific purpose other than appreciation: the viewer is allowed unique and uninterrupted commune with each image.

The book's introduction by anthropologist Hugh Brody examines the construct of photography in a general sense as well as the relationships between photographers and their subjects, particularly this photographer and his perceptions of landscape. Many of Brody's general comments are irresistible: "The watershed is vast, and developments thus far have come and gone or only touched its edges. But the very wildness of the place excites the frontier mentality. No one who works in or with the land can be unaware of ideas that encourage a sense that all real wilderness is doomed … Deep inside ourselves we all know that human spirituality is rooted in the earth and in values that we have come to associate with a life close to the earth: intuitiveness as a habit of mind, generosity and equality as habits of society ... Without a relationship to earth, open space, without a life at least partially shaped by the natural world, we confine ourselves to the urban, material and rational. In so doing we live without one half of what it means to be human."

Callbreath Barrington Hyland Helveker Shakes
Dodjatin Tsikhini Yeheniko Glenora
Clearwater–Chutine and Muddy Mess Creek

Checking out tributaries and picking up history in the summer of 1992, a contingent of FOS directors finally enjoyed their long-envisioned trip north to Telegraph Creek, aided by the grant from Shell's Environmental Fund. Travelling by automobile, Alaska State Ferry, and riverboat, Karen Hodson, Jon Shemerdiak, John Christian, Corinne Scott, May Murray, and her daughter, Joan Dunn, enjoyed a multi-day visit to familiarize themselves with some of the people and places referenced in our newsletters. Flightseeing trips by Ron Janzen of TelAir and riverboat excursions by Ron Ellis helped acquaint the southern river lovers with Stikine's physical realities while the locals—Dan and Diane at the RiverSong, Tony and Doreen at Red Goat Lodge, David and Tannis of Glenora—warmly introduced social aspects. Of course, the itinerary also included a soak at Chief Shakes Hot Spring before the return to more familiar territory ... and to another newsletter.

The swishing and swirling of incoming and outgoing information continued apace at FOS and our newsletter continued sharing its usual mix of items from near and far. One item raising concern at the time was the Stikine Watershed being omitted (dropped?) from the province-wide Protected Areas Strategy (PAS)—a single, integrated process for coordinating all of BC's protected area programs and objectives with the aim of identifying representative ecosystems for ecological diversity and recreation features while addressing needs and concerns of local communities. At its September 1992 AGM, the CPAWS-BC chapter passed resolutions addressed to the provincial government, urging the inclusion of the entire Stikine Watershed in PAS and in the CORE process (Commission on Resource and Environment). The rationale for Stikine's omission from the process was never determined and it was too late to worry about it.

Co-chair May Murray brought news of similar concern when reporting back from her Spatsizi hiking trip of late-summer, subsequent to the gathering of directors at Telegraph Creek. She was disturbed by a paper questionnaire being distributed by BC Parks volunteers at Cold Fish Lake. While the ministry's interest in beginning a planning process was commendable, she did share her concern about possible pro-development and mechanization bias in the questions, FOS, with known interest, had not received an original questionnaire. More alarmingly, for the first time in several extended visits, May reported seeing *no* wildlife during her multi-week ramble in Spatsizi.

Meanwhile, FOS Director Paul Christensen laid a critical eye on the hovercraft issue then gaining interest—In response to the commencement of hovercraft operations on the Stikine and Iskut rivers (July 1990), DFO had initiated an environmental screening of Cominco's use. The subsequent evaluation was conducted by the federally mandated environmental assessment and review process (EARP) which addressed three things: potential impacts

to fish and fish habitats; potential other environmental and related socioeconomic impacts; and public concerns. While addressing each article of the evaluation, Paul's comprehensive review did suggest growing skepticism around government and corporate environmental assessments. Edited for length, his report is here summarized with my italics added.

> Cominco's hovercraft travels from Wrangell to the Snip mine (60 kms) and back two or three times per day, five days per week. Its journey courses directly through "important migration, spawning, incubation, and rearing habitat for all five species of Pacific salmon, as well as for steelhead, cutthroat trout, Dolly Varden char, and mountain whitefish." Assessing potential effects on fish and fish habitat, three categories were identified: A1) noise, vibration, and shadows; A2) prop-wash and bank erosion; A3) increased risk of spills. In terms of item A1, EARP concluded that although the hovercraft may elicit startle and flight responses and could cause avoidance behavior in salmon when passing over spawning beds, these effects would *likely* be temporary. Also, because spawning beds are *primarily* in clear tributaries, back-waters, and sloughs, they are *relatively* unaffected by any such disturbances. A1 done. Since hovercraft ride a cushion of air, they tend to generate little wake (similar to that of small boats already operating on the rivers) and, therefore, EARP concluded that the propwash effects of hovercraft operations "did not contribute *significantly* to streambank erosion." So much for item A2. Recognizing the purpose of Cominco's hovercraft in delivering reagents and fuels in return for loads of gold-rich pyrite concentrate, EARP concluded the hovercraft "naturally increases the risk that deleterious substances may be spilled into the Iskut and Stikine Rivers." Surprisingly, the company (the operator) had yet to provide a contingency plan describing how it intended to prevent, contain, and clean up spills. EARP did not comment on the potential danger to fish or fish habitat in the event of a spill. So much for item A3.

> In assessing *probable* effects of the hovercraft on local wildlife populations, EARP ended its evaluation by countering, "There is considerable evidence that wild animals eventually become habituated to intermittent, loud airborne noises." All other potential impacts of hovercraft operation were either summarily dismissed as acceptable, negligible, or simply "not applicable" while some were discussed for the sake of discussion,

particularly that section regarding the commercial fishery. Similarly, acknowledged but unaddressed, were public concerns expressed through Skeena MP Jim Fulton which also included regard for international efforts at environmental protection as well as risks imposed by the mine itself.

Throughout the assessment, EARP suggested various mitigation measures to deal with potential environmental and other impacts. For the most part, however, they were all alike. Basically, the recommendations were: a) Cominco to choose an optimal route for their hovercraft operations to "maximize safety and cost-effectiveness while minimizing disturbance to fish and wildlife, to the commercial and Native food fisheries. The optimal route may vary with time of year … (blah, blah, blah); b) Cominco to cooperate with *concerned* agencies in Canada and in the USA "to develop environmental operating conditions which would *minimize* disturbance to fish, wildlife, and the fishery." Quite a lot of words with very little substance.

To the cynical eye, it seemed killing fish was okay providing they weren't being overly disturbed. Our hovercraft concern level was on the rise.

Before year 1992 came to a close, FOS welcomed long-standing member Tony Shaw to our board of directors. Having taught school in Iskut and having operated nearby Red Goat Lodge with his partner, Doreen, Tony came to the table with considerable knowledge of the subject watershed. (As an expert paddler and a fine instructor, he would also offer me a friendly introduction to proper techniques for playing in white water.)

In Memory of May Murray

Near the close of 1992, May Murray was feted by CPAWS–BC at its benefit dinner honouring her more than thirty years of dedication to environmental organizations including her roles as a founding member of both CPAWS–BC and FOS. Her dedication never wavered. She is shown here aboard *Empress of Stikine* on June 4th, 1986—photograph by Gary Fiegehen.

May passed peacefully in October of 2020.

Ever Onward

The year 1993 got off to a good start, sparing us new concerns while introducing news from the brighter side. On June 22, 1993, the Office of the Premier announced establishment of the 958,000-hectare Tatshenshini–Alsek Provincial Park in the far northwest corner of the province, abutting Kluane National Park in Yukon. On behalf of WWF, Prince Philip complimented Premier Mike Harcourt on his "far-sighted vision" while Parks Minister John Cashore celebrated it as part of an immense international protected area that will remain important for many years to come. Much credit must go to Johnny Mikes of Canadian River Expeditions and to Ken Madsen of Friends of Yukon Rivers whose early concern and tireless efforts were instrumental in raising awareness that was picked up by dedicated volunteers far and wide. Well done! Congratulations went to Premier Harcourt and his ministers for a courageous decision of global significance.

Taking another step forward, the BC Outdoor Recreation Council completed its year-long survey to identify which provincial rivers were deemed most threatened. The list was topped by Tatshenshini, followed by Fraser, Stikine, and Nechako. Though all of valid concern, this list could have easily been reversed by the time of its compilation. Nechako, once the major salmon supplier to Fraser was being threatened by a Kemano II addition, which would draw the river down a further twelve per cent. Stikine still carried mining initiatives on vulnerable headwaters, BC Hydro flooding reserves mid-river, and hovercraft chaos amid indecision on the lower river. Fraser was at long last being well attended to as a salmon source and never-to-be-dammed working river. Tatshenshini was, except for its very headwaters, now included in a provincial park with whatever protection that might offer.

A Welcome Launch

An invitation from Mary Ellen Cuthbertson to join her group of friends for a paddle on the lower Stikine was exactly what the doctor had ordered; better to risk drowning in the river than becoming smothered by its paperwork. Though helicopters and airplanes had provided good looks in the interim, it had been too many years (eleven already!) since I'd dipped a paddle in the lower river. With canoe *Dimples* strapped on top, it was a pleasure to drive north again in the Silver Dart pickup truck and a great pleasure to bypass Smithers and Dease Lake where four years of meetings had been taking place.

Although occasionally in paper contact, this was the first in-person meeting with Ms. Cuthbertson, then of the Southeast Alaska Conservation Council, since our 1985 downriver boat ride with Dan and Dave after the Telegraph Creek assembly. This time around, *Dimples* and I were to provide logistical support and a semblance of guidance for her and three river lovers from Washington State who would be paddling two by two in canvas-covered kayaks. Lindsay Smyth, a semi-familiar semi-local TCreek resident, offered to come along as an assistant guide and well-armed bear defender while also adding paddling power to our freight canoe. Off we went, now a party of six, here captured in a photographic selfie with Kate's Needle in the background: from left to right, Alison Peters, Lindsay Smyth, Peter Rowlands, Fran Partridge, Mark Sullivan, and Mary Ellen Cuthbertson.

Any journey log that may have been kept has long since gone the way of much memory: a surviving collection of photographs depicts very good weather and extremely low water—perhaps late summer. The weather held and the often-elusive lower Stikine Valley disclosed itself in the same cloudless brilliance as experienced during the 1986 *Empress* flight, allowing familiar mountain names to morph from the map into real-life river views. Especially welcome were clear looks at Commander, Cornice, and Cirque along with Saddlehorn, Devil's Elbow, and Eagle Crag.

In contrast to previous lower river paddling adventures, this one offered expansive campsites on sandy gravel bars at the mouth of most tributary streams, some of which offered landlocked pools with "reasonably" warm water for bathing. The luxury of an unhurried schedule allowed for short hiking excursions inland and nightly campfires on broad beaches, as well as the inspiration to indulge in poetic appreciation for the river and its tributaries:

> *Slowing and smoothing in softening descent*
> *Stretching and striding in sleeking long reaches*
> *Flickering and fluttering through Little Canyon*
> *Braiding acres of gravel across Scatter-ass Flats*
> *Waxing and widening, waving and welcoming*

> **Dokdaon Donnaker Missusjay and Jacksons**
> **Christina Patmore Oksa Deeker Jonquette**
> **Porcupine Scud Anuk Sterling Choquette**

> *Glistening and gliding through broad, tawny meadows*
> *Mellowing and meandering through forests of green*
> *Sneaking and snooping into backwater sloughs*
> *Prowling and prying, pressing and probing*
> *Seeking passage through a wall of granite and ice*

While enjoying two nights at MOF's old and improved Great Glacier Recreation Site (now a provincial park), we met a charitable solo paddler who had portaged his canoe into Great Glacier Lake for all to enjoy up-close appreciation of icebergs and a semi-close look at the ice wall. Wonderful. Joined by our new friend, here known as "Fritz" (his real name lost to history), we were soon back on the river for a sighting of Cominco's hovercraft heading upstream at high speed, followed by a look at Great Glacier Salmon's greatly enlarged and less visually appealing facility. Along with its relatively new bunkhouse accommodations, the site now exhibited sturdy docking facilities for the hovercraft and its Wrangell-based barge along with a sizeable crane for hoisting bags of concentrate to and from their on-shore storage area. Hmm.

Saying goodbye to Stikine's benevolent current, we semi-hardy souls (Fritz included) paddled the calm waters of Keteli Slough and made our way to Chief Shakes Hot Springs where mechanical disturbances of the main river were soon forgotten. Friends of Mary Ellen from Wrangell were there to meet us with their family of good cheer and a large box of refreshments. Salmon were spawning in the creek and paddlers were soon soaking in the tub. For the first time in my limited experience, the water level of The Great River's lower basin was dry enough to allow camping on the adjacent flood-plain meadow. Could it get any better?

When the four kayakers departed the hot springs in a motorboat bound for Wrangell and a charter flight back to TCreek, we three canoeists did a right turn to paddle and pull our way up the milky way of sediment-laden Shakes River into Shakes Lake to be astounded by the view: iridescent ice bergs on an aquamarine sea backdropped by the tri-horned spires of Castle Mountain. From our creek-mouth camp at head of the lake, the next day gave us an uphill hike to remember—a thousand-metre scramble through sparse vegetation and rocky ridges to the creek's uppermost waterfall, about a hundred metres below the edge of the icecap; we would have needed more time and equipment to go any higher. As it was, we had attained a fine place of rest with magnificent views all around. Insert all known superlatives for a high point in my existence—a speck of life afloat on a rolling sea of granite, one with the mountains and totally consumed by them—immersed in the magnificence that had attracted me from high-altitude some twenty years prior. Descending immense slabs of incredibly smooth and surprisingly steep granite with fading limbs and fading light reconnected me with the wisdom of a bum-skooch or two. Awe and gratitude remain.

The Elements

Events of the following year (1994) increased my awe factor along with gratitude for being absent. "Could it get any better?" My 1993 question while camping in the meadow at Chief Shakes Hot Spring was drowned out by the following year's record high water levels in the lower basin of River Stikine where camping would have been then impossible. Several days of very heavy rainfall that September saw the river-level drop rapidly after having gradually risen … much of the precipitation being captured as snowpack on the mountains. Finally moving inland with continuing wind and rain, the strong low-pressure system brought a mass of extremely warm air onto the coast and quickly melted snowpacks old and new. The inevitable flood on the lower river was the worst in memory, breaching riverbanks and allowing fishing skiffs to motor through the forest. The storm was larger and lasted longer than anyone predicted. A short distance north on the coast, it brought tragedy.

When storm conditions had finally improved after days of heavy wind and rain, Fisheries Biologist Johnny Tashoots joined pilot Ron Janzen in Tel-Air's C185 to transport an overdue load of sockeye roe from nearby Tahltan Lake to Snettisham fish hatchery near Juneau, Alaska. After delivering the roe and taking on a load of fry bound for Tahltan Lake, C-FVZP disappeared into cloud soon after leaving Snettisham float base and was never seen again. As soon as weather conditions permitted, a search party found remains of the aircraft and its people on a snow-covered mountainside. Probably motivated by a warm-sector lull in the storm action, Ron and Johnny were caught by the worst possible weather in the worst possible terrain: unpredictable winds and thick cloud were givens; a heavy airplane and poor visibility were contributors.

Being familiar with such flight conditions brings a certain amount of anguish and being familiar with the people on board makes it much worse.

Many people lost good friends that day.

Moving On

Amid the normal round of functional responsibilities during the following year (1994) came a multi-day Canadian Heritage River System conference in Peterborough, Ontario, as well as resumption of LSMAC meetings. Yes, for reasons unknown, the letter M had been re-inserted into the title and LSAC returned to its roots as the Lower Stikine Management Advisory Committee—just in time to be sued by Cominco. It seems an increasing level of expressed concern had brought a SLAPP suit (Strategic Lawsuit Against Public Participation) by Cominco against FOS and LSMAC, perhaps invoked for *undermining* their environmentally benign air-freight image. The plot thickened.

Taking our concern about the hovercraft directly to Cominco early in 1994 had proven unproductive. Later, on December 16, 1994, Maggie joined me for a second meeting, this time with the Snip Mine manager also present. On this occasion, the Cominco execs seemed to soften their stance, so much so, it produced a letter of intent to withdraw its SLAPP action against us and LSMAC. With the legitimacy of our concerns finally made known, we were able to stop rattling the chains for attention and take a breather. The next move was up to Cominco.

The masthead of THE CURRENT in the Spring 1995 edition showed my return as FOS chairperson while bringing along another series of remarks from the chair with typical fervour and acerbity. In addressing BC Parks' need for immediate input to its interim management guidelines for Spatsizi and Edziza, it was impossible to avoid comparison between the apparent *urgency* of determining mining and logging options within our parks with the outward *complacency* about destruction of fish habitat being caused by a (probably unauthorized) hovercraft on one of the world's finest remaining wild salmon rivers. Perhaps these suddenly urgent requests were part of a diversionary smokescreen to mask concerted attempts to quickly obtain the resources of the northwest before several recently enacted protective strategies kicked in: Forest Practices Code, Protected Areas Strategy, Biodiversity Strategy, et al. Having seen no change nor heard from Cominco for almost six months, it seemed obvious the company had adopted a business-as-usual approach to extract as much ore (profit) as possible out of the place before being distracted by external issues such as ours. While Cominco continued to be stalling for time in a state of total denial, our state of cautious optimism eroded away like so many metres of Iskut riverbank.

Although it was the obvious frontrunner, the hovercraft issue was not our only concern. In a letter to Premier Michael Harcourt and to Andrew Petter (Minister of Forests), the directors of FOS outlined concerns about the proposed four-fold increase in the annual allowable cut (AAC) for Cassiar District—from 140,000 to 400,000 cubic metres for each of the next seventy years. Citing a deficient basis of information and poor public consultation, FOS declared the proposed increase to be disappointing, irresponsible, and alarming. The crux of our position was the unjustified need to supply a greedy (largely external) pulp industry while ignoring wilderness values and the wishes of people who live there.

> We believe there is a place for planned, controlled, managed use of the Cassiar, and we would like to be part of a group that suggests some of the plans, controls, and management methods. The fact that the forest service district office in Dease Lake seems unable to fairly include the public in its plans for cutting the area should also be of major concern to you both. We do not believe it is the intention of the Government of British Columbia to disregard the public in such a manner and ask that more lead time be given to the public process of making land use decisions affecting the Cassiar Timber Supply Area. Finally, we believe it is fundamentally unethical to make major land use decisions and to sell timber licences (and other such commitments) before land claims negotiations have occurred.

With the diligence of Maggie Paquet, our directors' comments to the Cassiar Forest District directly addressed the proposed increase to the AAC in considerable detail. While highlighting inconsistencies in the scientific base, along with its too many unknowns, it was suggested a rush to market was trampling all other (mostly unassessed) values across local and provincial landscapes while ignoring global concern for the rapidly increasing rate of destruction of the boreal forests, a vital component of Earth's balanced ecosphere. "We believe the rush to log the Cassiar represents a last resort of logging companies that have been laying waste to the temperate zone forests south of the Cassiar district and to forests of Manitoba, Saskatchewan, and Alberta. If the clearcuts along Highway 37 in the North Kalum supply block north of Meziadin are any indication of the forest practices approved by the Cassiar District office, then we have great cause for concern." In closing, FOS recommended formation of a multi-stakeholder body to analyze resources and values across a provincial spectrum; suspension of all area logging pending completion of such analysis; and increased commitment to broader public input in land use planning for this large and diverse area.

There were other blips of interest on our scope. While Cominco's SLAPP suit seemed to have been defused for the lower Iskut–Stikine, there was now talk of a hovercraft being proposed for use by another mine on the nearby Taku River. Also, while recreation areas in Stikine Country were facing proposals for increased industrial use, plans became known for a proposed resource access road from Highway 37 to the headwater area of Mess Creek on the west side of Mt. Edziza; meanwhile, recently available aerial photographs were showing massive clearcuts abutting the southeast boundary of that provincial park without buffer strips on lakeshores and riverbanks. Nearby to the south, in order to access their multi-metals property adjacent to Cominco's Snip Mine and to their own at Johnny Mountain, Skyline Gold Corporation was contemplating extension of the Iskut Valley road beyond Eskay Creek while also considering possible use of a hovercraft. The venerable Craig River route for a road to tidewater also remained a worthy contender, especially given Alaska had recently authorized $2.5 million for a feasibility study there and pledged $22 million toward possible construction while simultaneously touting recreational tourism as a possible source of maintenance income. With recent cancellation of its Kemano II project on the Nechako, BC Hydro was resurrecting plans for dams on the Iskut–Stikine to help power the surging mining interests.

Meanwhile, for the fifth consecutive year, Cominco's hovercraft was running footloose and fancy-free along the lower Iskut–Stikine River where fishers and other boaters were expressing increased concern about bank erosion and significant losses to fish habitat. Ironically, by installing a trans-shipment dock for Cominco's ore on its property (with or without government approval), Great Glacier Salmon had doubled the threat to Iskut River salmon by doubling the number of daily runs possible by this less-than-benign air cushion vehicle. Seeing no signs of Cominco restraining its enthusiasm for use of this machine, FOS made application to the West Coast Environmental Law Association for financial assistance from its Environmental Dispute Resolution Fund in order to hire a recognized fisheries expert to look into the issue. With a $5,500 grant coordinated by Patricia Houlihan of WCELA, Dr. Gordon Hartman accepted the task and made arrangements to visit the scene.

14
Raging and Roaring

On May 16, 1996, I attended the Provincial Court in Vancouver to swear before a justice of the peace charging Cominco and Prime Resources with violations under section 32 and section 35 of the *Fisheries Act*—Cominco had sold this mine to Prime Resources Group on April 30, 1996, one week prior to a published report by Dr. Hartman which concluded that, along with deterioration of riparian habitat, Cominco's Hoverfreighter was stranding thirty thousand or more salmon fry each year along its route. With me in court (any citizen in Canada is permitted to file private prosecutions for breach of the *Fisheries Act*) was Christopher Lemon, a young lawyer with Ferguson, Gifford & Company, who had been recruited by Ms. Houlihan and her colleagues at the West Coast Environmental Law Association to provide guidance and support.

Because this Vancouver justice of the peace refused to issue process, I was forced, on May 30th, to attend before a justice of the peace in Prince Rupert. Satisfied there were reasonable grounds upon which to proceed, this justice issued a summons on Cominco and Prime Resources requiring them to make a first appearance on June 18th in Prince Rupert. Unfortunately, the summons was sent by mistake to DFO, which failed to serve it on time, thus making the legal proceedings more protracted. Prior to the rescheduled first appearance of July 9th, our counsel, Mr. Lemon, was advised that the federal government would be exercising its discretion under the *Criminal Code* and was taking over the prosecution. Arrangements were made for the first appearance to occur in Vancouver.

On July 18, 1996, Mr. Lemon, Dr. Hartman, and I attended a meeting with the assigned prosecutor, Ms. Valerie Hartney, at the Criminal Division office in Vancouver where I learned the prosecutor's role was to remain impartial while enforcing the law toward a just outcome. There was no guarantee the Department of Justice would continue with any prosecution: Ms. Hartney would stay the charges if, in her opinion, she could not meet the high burden of proof required to succeed on a criminal-type prosecution *(Fisheries Act)*.

On July 23, 1996, hovercraft CH-COM ceased operations on the Iskut and Stikine Rivers.

Our legal counsel suggested Prime Resources had recognized reality and had withdrawn the hovercraft from service to avoid current prosecution while remaining under the threat of future prosecution. Lack of evidence was not the issue. Otherwise, it would have been an embarrassing situation for the federal government to have one of its departments (DOJ) prosecuting for violations of the *Fisheries Act* while another department (DFO) had done nothing to exercise its jurisdiction under the same act. The Department of Justice appeared to be serious about pursuing the prosecution.

The long-anticipated letter from Canada's Department of Justice was dated December 4, 1996, and signed by Robert Prior, head of its Criminal Law Section. Addressed to our counsel Christopher Lemon, it summarized the department's responsibility for weighing the probable outcomes of prosecution while also protecting the public interest. It then stated: "After careful review of these materials, we are of the opinion there is evidence of the destruction of fish and fish habitat. However, when viewed in context, including the undertaking of joint environmental studies by Cominco and the Department of Fisheries and Oceans, we have concluded there is no reasonable prospect of conviction, and the prosecution will be stayed." Reminding us that the hovercraft ceased operation soon after federal intervention, the letter concluded: "The right of a private citizen to institute a prosecution for a breach of the law is a fundamental part of our criminal justice system. Your participation in this process is appreciated."

A letter from Canada's Department of Fisheries and Oceans was dated January 15, 1997, and signed by Louis Tousignant, Director-General of the Pacific Region. Similarly addressed to counsel Christopher Lemon, it expressed thanks to his clients and colleagues for their efforts in working to protect the Iskut and Stikine fisheries: "It is our view that it was primarily the evidence of fish destruction your clients obtained in early 1996 which led to the cessation of hovercraft operations." While agreeing with the federal attorney general's decision to stay charges, DFO concluded: "Please express our appreciation to your clients for their long-standing efforts which have resulted in a more environmentally friendly means of transportation for this mining operation."

Although the outcome of this legal initiative was predictable and somewhat understandable in the big picture, a degree of unrest remained … and still remains. Some of us, including Dr. Hartman, remained gravely concerned that in six seasons of operation hovercraft CH–COM had had a devastating effect on the wild salmon populations of the lower Stikine–Iskut. Information on state of the Iskut fishery (and the actual fish) is so far unavailable from DFO. Cominco took the money and ran. Prime was left holding the bag—and the mine. Loss of the hovercraft operation probably caused a decrease in its future profits, but that's it. No one accepted responsibility for the mismanagement. No one suffered economically. The citizens pursuing the issue were volunteers to begin with, and industry was never asked for remedial measures. For some of us, the issues created by the hovercraft remained front and centre long after its departure.

Dated April 15, 1997, a letter from Friends of the Stikine to the president and CEO of Prime Resources Group, and to the vice president of Canadian Mines–Cominco Metals, requested their companies to jointly pay $25 million restitution for destruction of fish and fish habitat in the lower Iskut River. Although the evidence collected by Dr. Hartman had terminated use of the machine, the *Fisheries Act* would not bring the salmon back to spawn, nor would the never-completed studies ever restore the fish and the river's riparian habitat. If received, the donated funds were to be placed in a Stikine Watershed Trust Fund (SWTF) to be managed jointly by Friends of the Stikine Society and the Tahltan First Nation with a mandate to restore riparian habitat and to protect wild salmon in the Iskut–Stikine River system. By investing a small portion of profit in the future of salmon, these companies could demonstrate their commitment to watershed integrity and ecological sustainability. Twenty-five years later, FOS has yet to hear back from Cominco Metals or Prime Resources. No worries here: perhaps they have found a better way to contribute. Patience and humour remain paramount.

At the time, this somewhat-tongue-in-cheek initiative was seen as a form of closure. Ideal as the proposal might have seemed in the big picture, it had little, if any, chance of acceptance even though the sum of $25 million was but a drop in the corporation's profit bucket. After all, beneath the paperwork, miners were simply doing what miners do to make money and keep their big wheels turning. Concerned citizens were just doing what was necessary to help maintain a healthy working environment, doing the due diligence that was neglected by paid government agents and wealthy industrialists. I am no longer a naïve idealist.

Cominco and Prime refused to consider any alteration of their profit margins. The federal government was complicit from the get-go and could not extricate itself from the tentacles of corporate control—DFO spokespeople always express concern for "the fishery" but never the fish. Unfortunately, given present-day state of the Stikine's wild fish stocks, it's difficult to see a happy balance or to accept closure. We'll never know the extent of the hovercraft's effect on the lower Iskut River because we hadn't determined what was there in the first place. The hatchery-enhanced stocks being maintained ever since in the Tahltan and Tuya tributaries are better than nothing, but it's disappointing to see healthy populations of all five Pacific salmon species being supplanted by "wild" fish farms. So far, salmon enhancement on the Stikine has only cost one airplane and two human lives … 1994 seems a long time ago.

The mining issue is an eternal conundrum. A primal first activity of humankind, it probably gave rise to everything under our current definition of industry and has now itself become one of the greatest threats to planetary health. The holes in the ground and the tunnels underground are not a problem—we want some of the goodies and need some of the others—it's our lack of regard for treatment and disposal of the residue that raises the red flags and demands immediate attention. Certainly, the cleansing power of Mother Nature is sufficient to absorb the chemical side effects of smaller mine sites—the loss of drinking water in one faraway village is not seen as a big problem in the global marketplace. However, given the number of mines worldwide and the indifference shown toward their re-naturalization (bonds and fines are merely the cost of doing business), the capability of our oceanic blood supply to handle the accumulated contaminants flowing downstream is a legitimate concern. With collateral destruction of fish and other water-bound creatures, the effects of mining extend far beyond any hole in the ground. Mining is inevitable; however, mining with ecological awareness seems a faint dream.

While we seem to understand the need for healthy blood in our bodies, it remains to be seen if, globally, we will ever appreciate the value of fresh water much as we do gold currency. Our economy runs on money, but our ecology runs on fresh water. The health of our entire planet depends on the health of our one common ocean, not on the size of the GDP or the GNP.

On the subject, the Red Chris property—on an eastern spur of Todagin Mountain above Ealue Lake—was already gaining attention of the local Cassiar Forest Watchers who first reported concern for Coyote Creek, affected by construction of the mine's access road. And, of course, there was talk of resurrection and completion of the adjacent railway line just over the hill. First staked in the 1950s, the Red Chris Mine began serious operation about sixty years later, digging for copper and gold concentrate to be shipped overseas from Stewart beginning in 2015. At time of this writing, the mine's majority owner, Imperial Metals of Vancouver, has contributed to construction of BC Hydro's Northern Transmission Line, which now connects Iskut Village to the provincial power grid along with this and other mine sites. On the downside, after perfunctory environmental review, the Red Chris Mine owners co-opted lakes in an upper tributary of the Iskut River as dumping ponds for their mine tailings. Ramifications are significant, especially in light of the 2014 dam breach in the tailings pond facility at Imperial Metal's Polley Mountain mine site which introduced concentrations of suspect chemicals into the natural purity of Quesnel Lake. While regulations have again been tightened after the fact, the effect of Red Chris on Todagin Mountain is a story yet to be told.

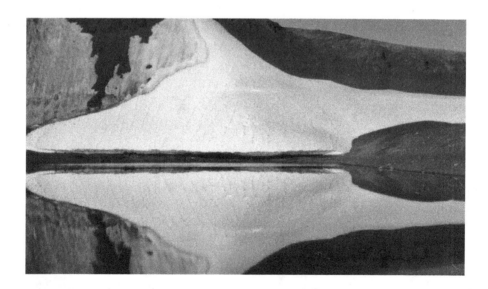

15
Pushing and Probing

Meanwhile, back in 1996, there were issues other than mining running through the Stikine story. As reported in our (FOS) October newsletter, there existed possibility for resurrection of good old LSMAC, which had not seen the light of day since being placed under the wing of the Prince Rupert Region's Inter-Agency Management Committee (IAMC) which is, in effect, the regional managers of provincial agencies. SLAPP suits and hovercraft excitement were probably major factors in the original move to ground LSMAC. One of the IAMC's most recent moves, *because of mineral access concerns,* was to remove everything west of Highway 37 from recommendations of the Regional Protected Areas Team (RPAT) which had recently completed four years of evaluating natural heritage values within the Skeena Region. An alarm was thereby sounded for the future of all river systems in northwestern British Columbia.

Another masterful piece by Maggie Paquet in the Winter 96/97 issue of *THE CURRENT* included reports on several relevant perimeter issues while bringing readers up to date with the immediate activities of FOS. Although the hovercraft issue had been put to rest, we expressed concern about the possibility of a terrible precedent being set for industrial activity in one of our healthiest remaining salmon producing river systems. The "Chair Marks" column had this to say:

> With removal of the entire northwest from the Protected Area Strategy in favour of mineral access concerns, and with the recently announced withdrawal of most provincial government agencies from Dease Lake, it appears streamlining is well underway with resource extraction industries poised for a field day on the open, not so level, playing field of northwestern British Columbia. If our justice system fails to send a signal to those who have allowed wanton destruction of our fish and fish habitat, then there is little hope of ever getting the message into the heads of our foreign-controlled

industrial corporations. While government agencies and organizations disintegrate for lack of money, the mining industry is tallying record profits.

FYI: <u>Net</u> income for BC mines in 1994 was $200 million, the second highest return in ten years. Solid mineral production reached a record high gross of $3.48 billion in 1995. Exploration investment had increased from $66 million in 1993 to over $88 million in 1995. Prime Resources, past partner with Cominco in the hovercraft operation and now sole owner of the Snip gold mine, recorded a $20 million profit for <u>one (1) quarter</u> of this year. And no one anywhere can afford biologists, baseline studies, or participant funding for watershed issues.

Prime Resources is a component of Homestake Canada, which is controlled by Homestake San Francisco.

FYI: According to the 1994 BC State of the Environment Report, there were 13 acid-generating mines in the province that have produced 72,000,000 tons of acidic tailings and 250,000,000 tons of acidic waste rock … <u>increasing</u> at a rate of 25,000,000 tons per year. *The Financial Post* estimated 2–5 billion dollars was needed to contain then-current levels of acid mine drainage in Canada. AMD is that unfortunate concentration of elements caused by mineral extraction that often finds its way to the bottom of the watershed and almost always proves detrimental to fish and other organisms. Promising as always to be good for the economy, a significant number of the 14 new mine proposals in BC were located in the Stikine Watershed.

Meanwhile, way out there in Ottawa, an apparent partial dismemberment of Parks Canada was accompanied by a major (about fifty per cent) reduction in the annual budget of the Canadian Heritage Rivers System, an action that renewed the call to action from certain corners of BC. While the concept had been on the table for two decades, the campaign seems to have then reached epic proportions, at least in my mind. Perhaps lifetime appreciation for Canadian fur-trade history was most responsible for me wanting that West Coast river link.

"We cannot wait for LSMAC, IAMC, RPAT, LRMP, UREP, LUCO, or any other acronym to come to terms with industry and First Nations before we take small steps on behalf of the river and all it supports. If we do not honour the outstanding heritage significance of the river, we have little chance of ever getting around to respecting the ecological values of the watershed. CHRS designation will not protect the Stikine. It would, however, say much about our country and our commitment without complicating the protected areas debate." So went my last official rant on that subject.

"Thanks for your patience, friends," was the appropriate opening into our next newsletter, the Summer/Fall 1997 edition of *THE CURRENT*. It had been twelve months since Cominco/Prime's Hoverfreighter had been removed from service, and with a full year of paperwork and phone calls, and meetings and conferences augmented by occasional press releases and media interviews, there had only recently been time to enjoy procrastination. With it, however, came realization that much of the news was difficult to write about and some of it was too damn ugly to think about. As evidenced by the hyper-acerbic tone of my "Splinters from the Chair" report for that newsletter, fatigue and frustration were kicking in. Some excerpts:

> In case you are experiencing difficulty in connecting all the dots, it goes something like this: After permitting six seasons of unmonitored hovercraft operation, DFO now warns PRIME not to operate the hovercraft because of evidence of fish destruction by COMINCO/PRIME, neither of whom can be prosecuted because COMINCO had once initiated never-completed environmental studies with DFO just before selling their share of the mine to their partners in PRIME, as in Prime Resources Group Inc., an employee-free profitable shingle for Homestake Canada which is controlled by Homestake San Francisco who now manage the SNIP mine for its new owner, Prime Resources Group Inc. Some slick, eh?

> NB: The so-called DFO–Cominco joint environmental study of 1995–96 was never completed because neither DFO nor Cominco would send their respective biologist into the area of study (the Iskut River) except by riding on the hovercraft. *[The fact of this joint study was instrumental in excusing Cominco/Prime from prosecution.]*

> While some people consider it a legal victory for the hovercraft to be removed from service, there was little celebration for those friends heading off to LSMAC, LRMP, and other public participation forums. While waiting and praying for the river and its salmon to recover, it was difficult not to consider Bronson Slope and Red Chris mine developments proposed for the Iskut River and the rumoured-to-be-acidic holding ponds at Gulf Canada's coal mine on the upper Spatsizi and Klappan.

Another Acronym

After a lengthy hiatus, LSMAC reconvened on December 3, 1996, to confirm its resurrection and to announce its entry into the world of LRMP—the Cassiar-Iskut-Stikine Land and Resource Management Plan—the high-priority first of three such plans for northwestern British Columbia (also Dease–Liard and Atlin–Taku). This Cassiar–Iskut–Stikine LRMP encompassed the entire Iskut–Stikine Watershed (Canada Limited) and included the Canadian portion of the Unuk River system to the south as well as the village of Dease Lake on the northern divide.

Although the likely participation of the Tahltan Tribal Council in the LRMP process suggested the possibility of "doing right" in terms of social justice and ecological sanity, I was less than enthusiastic about overall results. Perhaps too many LSMAC meetings going nowhere had taken their toll—being reminded of the panel's importance while again being told how little time and money were available to make it happen ... a room full of well-meaning people overwhelmed by maps, figures, and documents while heartfelt expressions of love for the country get shuffled down the table and up onto the side of a flip chart before eventually falling off the end of the agenda along with other intangibles.

The following LSMAC meeting on April 8, 1997, introduced Marlin Murphy, environmental coordinator for Homestake Canada at the Eskay Creek and Snip mines, who confirmed the hovercraft had been removed from service and was up for sale. Ignoring expert advice regarding salmon fry vulnerability and ignoring any requirement for legal authorization (DFO, DOJ, et al.), Hovercraft CH-COM was later taken out to Wrangell before June high water. (Several months later, negative effects on fish and fish habitat were raising concern about a recently employed AP1-88 hovercraft delivering mail and freight to eight Indigenous villages along the Kuskokwim River in western Alaska.)

Also present at this LSMAC meeting was Pierre Lemieux of DFO Smithers who was unable to answer many of our questions because the Stikine is outside his jurisdiction. However, he was able to confirm that the Canada–USA Stikine Salmon Enhancement Program, administered from Whitehorse, had introduced millions and millions of hatchery-enhanced salmon fry to the Stikine system in each of the previous six years. As well as enhancing the Tahltan Lake sockeye run, DFO also began out-planting (seeding or salting) Tuya Lake with sockeye fry that had also been raised in the Alaskan hatchery from eggs collected at Tahltan Lake. Because a velocity block on the Tuya River prevented the upstream return of spawning fish, this was a one-way run harvested in the Stikine mainstem as part of the combined and undifferentiated Stikine–Iskut fishery. In 1990, the year the hovercraft entered service, DFO tripled the previous year's number by introducing 3.6 million fry to the Tahltan system. The Tuya system came "onstream" the following year with approximately one-half of the annual allotment, which by 1995 had increased to five million fry. The fact that three million fry were introduced in 1991 and four million in 1992 might explain why the (four-year cycle) sockeye returns of 1995 and 1996 were more than three times the pre-hovercraft annual average.

Men occasionally stumble over the truth, but most of them pick themselves up and hurry off as if nothing happened.

This quote from Winston Churchill opened a report by Jenny Klassen and Maggie Paquet on the introductory workshop of the Cassiar–Iskut–Stikine LRMP held in Dease Lake on February 27 through March 1, 1997. Speaking on behalf of the Tahltan Joint Councils (Telegraph Creek, Iskut, and Dease Lake), Yvonne Tashoots, chief of the Tahltan Band Council, reminded all participants that the Stikine–Iskut Watershed is more than a land management unit for her people—it's a way of life. As such, the Tahltans became the initial First Nation to participate in a provincial LRMP.

At the subsequent first meeting (April 4–5), when asking about roadless zones, participants were advised that such things were not within the scope of an LRMP. Huh? Concern about the limited time frame for the process (two years) invoked requests for a moratorium on development for duration of the process in hopes of avoiding a business-as-usual scenario as experienced elsewhere. "No way!" declared the government representatives, citing such an expectation as being unrealistic. The best part of the day was an outstanding performance by the Tahltan Youth Dancers, accompanied by voice and drum.

Two subsequent meetings in May established ground rules and terms of reference leading to creation of a steering committee and a communications subcommittee, both of which included our table member Maggie with Jenny and me remaining as alternates. The Technical Support Team consisted of government staff along with Glenda Ferris, process coordinator for the Tahltan Joint Councils. Stuart Gale was facilitator of the proceedings, and Tom Soehl, operating under direction of the Prince Rupert IAMC, was the process coordinator. Despite a series of newsletters and a website presence for the process, many of us remained uncertain about the LRMP's true essence. Maggie put it this way:

> In spite of all the apparent communication in this LRMP, some of us are still wondering what LRMP means. Could it be: Let Regressive Miners Plunder? Logging Really Magnificent Parks? Large Reclusive Mammals Pulverized (by increased access and habitat loss)? Judging from what has happened elsewhere in BC, it could be any or all of these. Let's say we are still skeptical. Many of us are bone tired and wondering if we're being used to validate a process that promulgates such an oxymoronic concept as *sustainable development*. After all, our planet has a finite carrying capacity. That's not rocket science. How much longer can we delude ourselves that we can keep on taking, using, changing—usually irrevocably—the landscapes and the gifts they contain?
>
> Economics is the skyhook upon which government and industry hang their agendas. If you can call it science, economics is at best a social science, and social means people. What are people, if not intimately connected to and evolved—culturally and biologically—out of the landscapes we inhabit? The landscapes for which this LRMP is known world-wide are large wilderness landscapes. This is the wilderness out of which Tahltans have created a nation and a way of life.

Appreciation for Maggie's efforts prefaced my thanks to everyone, especially to our friends-of-the-river members who had been steadfast in their support during the previous year's particularly challenging agenda. While celebrating an entire year without hovercraft activity on the lower Iskut–Stikine, my "Rumours and Revelations" column showed Hoverfreighter CH–COM remaining front and centre in some minds:

> Information recently purchased under Freedom of Information suggests that good old hovercraft CH-COM may have operated illegally for its six-year tenure on the lower Iskut and Stikine rivers. It seems that when they went to sell this ACV, Cominco had no Certificate of Registration (C of R) to hand over. According to the mine manager at the time, this document must have disappeared in a small fire that occurred at the Bronson Creek facility in 1990. Ergo, according to our logic, this essential document could not have been onboard CH-COM, where it was required by law, during operations from 1991 to 1996. *[Journey logbooks and any record of loaded weights may have disappeared in that same small fire.]*
>
> Now that you-know-what is no longer you-know-where, an American air carrier (Southern Air Transport) is using American crews and American registered C-130 Hercules aircraft to airlift the Snip Mine gold concentrate from Bronson Creek BC to

an American port (Wrangell) for an American company (Homestake). Be not alarmed! The important point here is that the Snip gold mine is now being properly serviced by air access, as it was originally authorized. It appears to be a first-class airlift by people who know what they are doing; and we wish them good luck for a safe and efficient operation in very demanding conditions.

Ironically, if not hilariously, it seems anyone interested in watershed integrity has no money, and the only people making big money out of the Stikine are not interested in the subject.

Integrity indeed! While continuing to chastise industrialists for their apparent lack of concern for watershed integrity, I was beginning to lose personal integrity. While lobbying for ecological sanity on a platform of "we're all in this together" I had unconsciously embraced the insanity of picking sides. It didn't start out that way. My participation in the anti-dam activities of the 1980s was pure and simple: I had added my voice to the discussion, and regardless of the final result, it seemed important to throw one's hat into the ring, to take part in the dialogue, to be a pro-active citizen, to be part of the community. Should we have failed in that cause, we would nevertheless have shared in celebrating an outstanding feature of our natural world. Even in failure, a step might have been taken toward firmer ground. Meanwhile, we were introduced to the people who call Stikine home and received an overview of the region's history. For many of us, respect for the Tahltan people and their Indigenous land claim was the obvious starting point in any resource management discussion.

My intended retreat from the campaign trail did an abrupt U-turn when the hovercraft appeared on the lower Iskut–Stikine. This seemed like a no-brainer: the operators would take the hovercraft off the river as soon as they understood it to be more harmful than anticipated. Not so. "It's a big muddy river and you can't prove a thing." These words from a senior Cominco executive came as a shock, especially after having voluntarily shared our legitimate concerns.

With considerable naïveté, I was expecting to hear muted surprise followed by a sincere-sounding pledge to look into it. Not so. Notwithstanding humankind's endless embrace of industrial progress, how could we, in this day and age, continue to knowingly waste countless numbers of wild salmon in the name of corporate efficiency? Fresh water. Fish. I became hooked and couldn't let go. When the line tightened, I began to fight. As evidenced by the nature of my commentary on the subject, tensions were building inside and out. Only in retrospect do I appreciate the level of ACV-related stress occurring within government and within industry as well as between the two factions. In addition, while undergoing political realignments within itself, the Tahltan community was probably experiencing a high degree of angst in having to deal with enviros as well as with government and industry.

That "Rumours and Revelations" column was the last of my lengthy rants on any subject, and that Winter 1997–98 issue of *THE CURRENT* was the last newsletter to see me as FOS chair. A piece of sloppy writing in my report on the subject LRMP got me ejected from the LRMP table and simultaneously kicked out of Tahltan territory. A proof-reader had also missed seeing the danger, and last-minute haste and paste of construction had left the phrase "cowboys of the Stikine," referring to the Tahltan people in general, instead of to its industrial development faction as originally intended. In any case, it was a sloppy piece of writing for which I took full responsibility. Perhaps, on another occasion, my failed parody might have created less of a stir; however, these were tense times, and the pressure was on. The provincial government was being challenged to find a balance between industry and environment—their external angst and inter-agency tension was palpable. The Tahltan Tribal Council, the first-ever Indigenous entity to participate in any provincial LRMP, was under considerable pressure from within and without ... and then along came this mouthy enviro who was directly involved in the recent cessation of hovercraft operation that had removed paying jobs from Tahltan people.

From the outset, my time with Stikine had been an education. My report card shows a passing grade in elementary Earth sciences (geography, geology, and liquid dynamics) along with miserable failures in the more advanced human sciences (sociology, psychology, and group dynamics). My decision to remain involved in Stikine River affairs subsequent to resolution of the dam issue had been a conscious one: I wanted to know how land- and resource-management decisions were being made, to see how government works, to see what makes the province tick.

The Lower Stikine Management Advisory Committee opened that door for me with a "good news–bad news" agenda—a large number of knowledgeable and well-meaning people confronting piles of paperwork erupting from the depths of political administration. As evidenced by its inconsistent "management" title, the committee was a brave step forward into unfamiliar territory. For the newbie, it was heartening to watch LSMAC strive toward objectivity while riding the untamed horse of industrial priority. Despite its underlying framework and occasional back-door transactions, LSMAC was, in my opinion, a general success in its surpassing of tokenism to provide a good hard look at the lower river. Although clear-cut logging ran rampant elsewhere in the region during LSMAC's tenure, the forests of the lower watershed now appear mostly reserved for natural evolution in concert with small parks and protected areas. May it ever be thus.

The ensuing Stikine–Iskut LRMP process proved too much for me. Along with life's normal complexities had come six years of high-pressure hovercraft affairs on top of LSMAC participation being followed immediately by the Cassiar–Iskut–Stikine LRMP, which promised further high-anxiety frustration in lobbying for ecological ideals in an industry-biased forum. Wildness and wilderness seemed of no inherent value unless incorporated into a recreational tourism resource and subjected to cost–benefit analysis alongside logging and mining. The ministries of natural resources, federally and provincially, were administering to the extractive industries of these resources while having minimal concern for health of the resources proper—we only needed enough fish and trees to keep the industries healthy. Nothing seemed to have value unless it had financial value in the marketplace. Business had evolved from being part of life to being life itself. Though our government ministries are filled with good people, many possessing ecological awareness, all actions and decisions within the ministry are necessarily subject to its limited mandate. Nowhere in this lengthy process had the value of fresh water been given anything but lip service. Locally and globally, water remained largely undiscussed and unprotected while awaiting capture and containment as a marketable commodity. The question of shared rights to this precious resource remained unanswered... as did questions around Indigenous land issues. Meantime, big industry from somewhere else was selling us plastic bottles of drinking water from gas stations located adjacent to our freshwater wells.

Eventually, impatience and an overzealous ego awarded me a serious lapse in judgment. Learning that I didn't belong in the process was part of my education.

Of course, it hurts to have been banished from a place I will ever hold dear; however, that's the risk of doing business. It was no one's fault but my own: fatigue and frustration had teamed up with bullheaded bravado in helping me lose sight of subtle priorities and sensitivities in a climactic situation. If offered a second chance, I would retire from the scene well beforehand to simply watch the process unfold as it would. Sadly, I had allowed myself to become lost in the process of process—losing sight of the forest for the trees—and I carried on pushing the envelope without realizing how close we were to the edge. With a fine line existing between satire and sarcasm, there are times when humour should not be attempted, and this was one such instance. Some people will never forgive me while others still laugh along with my failure—fishers, hunters, miners, and ranchers, yes—cowboys, no.

"We are not the cowboys of the Stikine!" declared the confrontational assemblage of council members—customarily decked out in their cowboy boots, blue jeans, and plaid shirts while sporting big, shiny belt buckles, and cowboy hats. That colourful scene will always be with me, along with eternal regret for having insulted the people I most wanted to support. Friends of the Stikine was one of the first, if not the very first, provincial citizens' group to call for resolution of the Tahltan Aboriginal Land Claim. Apparently, while we (me) remained bullheaded on the issue with government and industry, the Tahltans were resolving the issue behind the scenes, in their own time and in their own way. Maybe some of us, particularly this one of us, should have gotten out of the way sooner.

Meantime, our whining and complaining about lack of enforcement of government regulations was having some effect. For the third time in eight years, Revenue Canada saw fit to audit Friends of the Stikine for activities deemed too political; a charitable society is allowed to be only ten per cent political. We were in danger of losing our charitable status. The fact that almost all our energy since the previous audit had gone into a certain transportation issue on the lower Iskut and Stikine rivers might suggest Friends of the Stikine had brought new meaning to the term *political*. If doing industry's environmental studies and government's environmental enforcement is deemed to be political, then our organization had indeed gone over the line. My last mini rant under the FOS banner concluded: "Though it's not impossible to make money in a police state, how is it possible to be political in a corporate state?"

Fortunately, saner heads rode in to save the day. Ric Careless and Gil Arnold were instrumental in saving our charitable status, though not our place at the LRMP table. Ann Jacob and Stan Tomandl came aboard to help May Murray keep the good ship FOS afloat while Maggie Paquet and I both resigned and moved off in different directions.

16
Settling and Surrendering

"Understand this about the Stikine: We are not just talking about a river here. We're talking about one of the most magnificent stretches of running water on the planet."

These words by Mark Hume from his book, *The Run of the River,* superbly capture the sentiment that had originally planted the seed and ignited my effort toward seeing Stikine included in the Canadian Heritage Rivers System program. Always prevalent in my mind, the idea had often been shuffled off to one side in favour of more stringent protective measures or had become lost in planning conversations and prolonged deliberations at the provincial level. When the Iskut became committed to industrial use—mining and probable hydroelectric installations—protection of the Stikine mainstem seemed more important than ever, not only for salmon and other creatures, but for ecological sanity in the global sense. We hoped CHRS status would provide a form of low-maintenance recognition. Having killed our greatest salmon river (Columbia) and spending millions of dollars per year to keep the second greatest on life support (Fraser), it was difficult to believe that, at no cost, we would not protect one of the world's last great Pacific salmon rivers. If we were to do more than pay lip service to biodiversity and sustainability, we would need to take immediate steps to protect the natural balance of the Stikine watershed. With the primary upper tributary and the primary lower tributary already given over to mine development, it seemed vitally important for the mainstem to be protected in hopes of maintaining the ecosystem's integrity.

A major step forward in recognition of our rivers occurred in 1995 when the Province of British Columbia established the British Columbia Heritage Rivers System (BCHRS), the first provincial system of its kind in Canada. Simultaneous with creation of its BCHRS, the Province of British Columbia finally, and at long last, joined the Canadian Heritage Rivers System program (CHRS), which had been created in 1984. "Hallelujah!" came the cry from many joyful river lovers. Although, BCHRS designation does not carry legal or regulatory power, similar to CHRS, it does provide greater emphasis on river-related values during land-use planning processes while reinforcing the overall importance of healthy river systems, helping protect a river's special values by encouraging greater public involvement in its stewardship.

Helped by leadership and financial support from the Ministry of Environment, Lands and Parks, and from the Ministry of Forests, a government-appointed board recommended twenty rivers to the BCHRS program between 1995 and 2000. Stikine was included in the inaugural batch of rivers along with Adams, Babine, Blackwater, Cowichan, Fraser, and Skagit. It had been a long road since Riverfest I in the early '80s to Riverfest V in 1996—meetings and gatherings, surveys and questionnaires mostly sponsored by the likes of Nora Layard, Robin Draper, and Bob Peart at the Outdoor Recreation Council of BC. Although details are covered in dust, it's a safe bet that the relentless energy of Mark Angelo, former chair of ORC's Rivers Committee, was instrumental in creation of the BCHRS where, appropriately, he became that board's inaugural chair. Of course, a letter from FOS in appreciation of his efforts also promoted the Stikine mainstem for nomination to the Canadian Heritage Rivers System which the province had simultaneously joined.

It was not to be, has yet to be, and may never be. As of this writing in 2022 hindsight, the Stikine River is not included in the CHRS, though perhaps it was one of the five originally nominated by the province when it joined the national program. Nevertheless, the Fraser River is a most worthy premier nominee; Cowichan and Kicking Horse are suitable companions in our province's representation; and BC's presumptive first choice Tatshenshini became a Yukon entry into the national program. My personal crusade for inclusion of Stikine in CHRS would temper with time until coming to rest in 2001 when the upper river's designated recreation area was upgraded to provincial park status, thereby adding a significant layer of anti-dam protection to the Grand Canyon of the Stikine. Respect for the river is more important than any classification we give it.

By the mid-1990s, any need for national park reserve protection for the Stikine River had been dispersed in the winds of change; and, although provincial park status would soon be applied to the upper mainstem, the source waters were still crying out for a *headwaters* reserve. For more than ten years, surrounded by mounting concern for ecological sanity, coal-mining interests had continued their evaluation of the Mt. Klappan deposit in the upper headwaters by drilling test holes in the ground and by seeking better financing in the board rooms. Although stocks changed hands and companies changed names, the physical mining process never did begin. With hundreds of millions of dollars at stake, the international mining consortium lobbied for construction of direct road access to the site as well as for the laying of track to activate the abandoned rail grade. Although an approved airstrip was in place alongside a test-camp facility that continued to grow in size, a fully functioning mine site had yet to evolve.

While Tahltan elders and others continued to oppose any such development in this part of their traditional territory, the government of British Columbia undertook environmental assessment of the overall project, thereby bringing broader perspectives to the issue. As a result, in early 2015, the provincial government paid $18.3 million to buy back the sixty-one coal mining licences in the area, thus initiating a permit deferral, pending completion of land and resource management negotiations with the Tahltan Central Council.

Although Fortune Minerals and its Korean-based partners remained eligible to reacquire these shares if coal mining was subsequently approved, there was, for the first time in a long time, some hope on the horizon for this extremely sensitive and ecologically valuable region. Happily, the regional LRMP of year 2000 provided special management zones for both the Spatsizi and Stikine headwaters. A welcome sigh of relief was heard.

Soon after the turn of the twenty-first century, another player came onto the scene. Royal Dutch Shell suddenly had rights to drill for coalbed methane throughout this same headwater region adjacent to Mt. Klappan. Fortunately, another voice joined the conversation about the same time: a group of Tahltan elders identifying themselves as Klabona Keepers became organized and dedicated to protection of this high-country section of their traditional territory, which they aptly named the Sacred Headwaters. Hallelujah! Then, fortunately, after a decade of intense lobbying and high-intensity negotiations, Shell's contentious fracking issue disappeared from the landscape as suddenly as it had appeared: in 2012, the government of British Columbia wisely banned oil and gas exploration in this headwater region while granting drilling rights to Shell elsewhere in the province. Heartfelt thanks go out to the Klabona Keepers for putting their hearts on the line. Congratulations go to the British Columbia Legislature for facilitating a no-net-loss decision, and kudos go to Royal Dutch Shell for demonstrating the down-home wisdom so often lacking in our international industrialists. Looking east across the ecologically intact Sacred Headwaters, a large glimmer of hope could be seen on the horizon.

Damnation

Reflecting on my Stikine experience, I sometimes wonder if my initial aversion to BC Hydro's Site Zed dam was gene-related, in the sense that both my parents had served alongside the RAF's #617 "Dambuster" Squadron in northeastern England during WWII, and at my first childhood solo visit to a movie theatre, *The Dam Busters* held my attention for a second-consecutive showing until my father found me in the dark and dragged me home for a late dinner. In those days, dams were okay: the one on the Avon River in Stratford, Ontario, gave us a lake for paddling and skating.

My perspective began to change with the first sights of big dams on big rivers such as the Columbia and Colorado. I am not an environmentalist and never did I intend to be one; however, it was impossible to ignore the Stikine–Iskut five-dam issue, and I joined the fray as a concerned citizen. As the campaign progressed, I began to realize the Stikine River was more capable of saving us than we were of saving it—the river can teach us much. To my mind, fresh running water is vital to our planet's well-being, and putting dams on wild rivers is equivalent to putting tourniquets on healthy arteries. How many arteries can we block and still keep the body alive? How many more dams can we insert into our ever-diminishing supply of fresh water before killing our common ocean? One too many is too many. By the time of this writing, there are sixty thousand large dams (more than fifteen metres in height) worldwide, with literally millions of smaller ones. On Canada's

8,500 named rivers, there now are fifteen thousand dams, 1,157 of which are classified as large, and 450 of which are purposed for hydroelectric generation.

Although the term "run-of-the-river" remains open to definition, several small-scale hydroelectric developments now on Stikine tributaries seem far healthier than the originally proposed five-dam megaproject with its enormous reservoirs, themselves far from benign. Small-scale hydroelectric generators installed high above fish habitat makes sense, especially if serving local needs; otherwise, we run the risk of being trapped in our own spider web of transmission lines. Newfound reluctance to impose further "enhancements" on the Nechako River also bodes well for the general health of BC's northwestern rivers. Given current state of the ocean, its fish, and our climate, it seems wise to refrain from putting another dam on any major river flowing to the Pacific Ocean.

After more than a decade on the back burner, the Site C dam proposed for the Peace River popped back into prominence in the mid-1990s ... and squelched any wild strains of cautious optimism I may have had. Although not on a Pacific-bound river, Site C is the third tourniquet being applied to Canada's largest freshwater artery. Hardly a run-of-the-river installation by my understanding, its recent construction delays suggest the original groundwork to have been faulty. No longer deemed unnecessary as it once was, several attempted resuscitations in the 1990s led to political re-approval by Premier Gordon Campbell in 2010 under the guise of environmentally friendly clean energy. In construction since 2016, while pushing the limits of fiscal responsibility, the Site C project has also redefined the term run-of-the-river—a nine-thousand-hectare reservoir behind a sixty-metre-high dam with spillways spanning the river.

Of special note: in the ten years since this project's approval, electricity consumption in British Columbia has remained far below BC Hydro's forecasts despite ever-increasing population, a situation similar to the Stikine-era miscalculations. However, notwithstanding possible effects to the health of its globally significant freshwater delta in the heart of Canada's largest national park (Wood Buffalo), a third dam on the Peace River might be better than a first dam on any one of our remaining wild rivers. The Peace ceased to be a free-flowing river with installation of the W.A.C. Bennett Dam in 1968, and its fate was sealed in 1980 with completion of the Peace Canyon Dam twenty-one kilometres downstream.

When it comes to the financial picture, I am reminded of a respected individual's assertion that the Stikine–Iskut five-dam megaproject was never ever going to be built—our efforts on behalf of the river had only upped the game's ante—because the entire project was merely a stock-leveraging, money-making ploy by the very rich and powerful. Perhaps my level of naïveté was even higher than I previously thought. Nevertheless, though our efforts may have been for naught, there is some comfort in knowing we were not alone in our concern. Besides, that dam issue did bring about my life's greatest airplane ride.

Back on the ground and back to the present, there seems to be peace in the valley. The Cassiar–Iskut–Stikine LRMP process was completed at turn of the century and appears to have been successful in its mission without Friends of the Stikine at the table. With special thanks to wisdom from the grass roots, now collectively known as Klabona Keepers Elders Society, the fragile headwaters of this major river system have been recognized for their value and have a degree of short-term protection against industrial desecration by their inclusion in a special management zone; however, with agreement of Tahltan First Nation and the government of British Columbia, mining could yet occur in the region and coal miners are still poking about.

Meantime, a 427-hectare provincial park contains the ultimate source water lake of Spatsizi River while designated protected areas (Chukachida, Pitman, etc.) adjoin the 257,000-hectare Stikine  River Provincial Park which extends down the mainstem to Mess Creek, encompassing the entire Grand Canyon and abutting Mount Edziza, Spatsizi, and Tatlatui provincial parks. Well done, British Columbia. With the Tahltan First Nation presumably involved in resource management, we can reasonably hope the inevitable mining initiatives in the area are to be conducted in such a scale and manner as to respect the health of the Stikine River.

Flowing between Mount Laura on the left and Kate's Needle on the right, Great River Stikine captures the natural majesty of place and evokes feelings of respect.

On the way out the LRMP door, Maggie's Winter 1997/98 edition of *THE CURRENT* gave us May Murray's report of the FOS-organized lower Stikine River rafting expedition which had taken place from August 27th through September 3rd, 1997. A principal feature of the trip was the opportunity to learn more about Tahltan culture; and to this end, permission was given by the Tahltan First Nation chief and councillors to begin the trip at Tahltan Flats with a performance  by the Tahltan Youth Dancers, together with a feast and a guided walk around that sacred place. The eleven rafters were to be accompanied there by Iskut residents Jim and Erma Bourquin, their sons, and Erma's father, Loveman, an elder of the Iskut Band.

The weather was terrible. Heavy rain forced the performance and feast to remain indoors at Dease Lake where baked sockeye, moose stew, salad, and bannock launched the expedition in nutritious fashion. With thanks to rafting guide Johnny Mikes, the river trip proved equally nutritious, and the weather became more reasonable (by lower-river standards). Aided by local families (Pakulas, Klassens, Sampsons, et al.), the nine-day excursion was a total success. Adventurous newcomers, adults and children alike, were impressed by the beauty and breadth of Stikine while extending personal boundaries. Without any need to portage, this downstream float on the lower *one-half* of the mainstem required fourteen allocated days incorporating four airplane rides and three power boats, in addition to the food and hardware required to maintain two inflatable rafts. Flowing strongly through harshly beautiful landscape, the lower Stikine River is seldom easy to navigate and often filled with surprises.

The truth of these words hit home for me while accompanying Tony Shaw on a guided field trip for outdoor education students from St. George's, a senior boys' prep school in Vancouver. It was May 2007, and having been seen in Tahltan territory several times since 1997 by Tahltan First Nation councillors without being apprehended, I assumed my banishment to have been political in nature and temporary in effect. Subsequent visits would seemingly support that supposition, though an official easement would never be expected. My respect for the Tahltans has never wavered. My most serious regret is not recognizing and accepting subtle invitations from some of them during my earliest days in their territory.

Amen

Having never lived in the same space for more than three years at any time in my life, Stikine had given me a sense of belonging—a first appreciation of *place*. With that realization had come subtle introduction, like a waft of woodsmoke on the river, to the other human presence and to a different way of relating with place. Interesting and complex. The connection Indigenous people have to their homeland is impossible for me to fully appreciate; however, such a connection does exist and is important in ways many do not understand. Despite our best intentions, painting the map green with protected areas does not necessarily respect that relationship. Fewer and fewer of us in this global village are still living on our "native" land, let alone attached to it. Waving a national flag is not the same … allegiances change, and pledges are but words. Being well aware of my good fortune and ever thankful for the lifestyle blessings of my social system, I nevertheless find myself more spiritually aligned with Indigenous peoples than with our worldly religions. In fleeing European depression and oppression, our settler society introduced many new things to the Indigenous peoples of North America. Wisdom was not one of them. Alas, similar to my Stikine experience, we colonizers, by duty or design, were too righteous for our own good, and may have missed an opportunity to receive more than we could imagine … despite human atrocities present in both cultures.

On the north coast of BC, the winter of 2006–07 had been huge. Snow was deep and spring was a long time coming. Fortunately, both our field trip's leader and the students' teacher were well informed and well prepared—Tony Shaw and Neil Piller were good friends and accomplished paddlers who had previously conducted similar outdoor adventures for students, an activity they considered well worthwhile. Tony and Neil would continue to challenge themselves together in bigger water elsewhere. Meantime, I was being invited to help keep Tony's boat aligned with the current while he and Neil kept their eyes on "the kids"—eleven high-spirited teenaged males escaping the classroom but having had limited experience in small boats and big water. It was far more fun than anticipated and the weather was far better than expected.

Carried as an inconvenience for the initial two days, our compulsory snowshoes were a welcome necessity for remainder of the expedition where we often camped on snow with uphill day hikes into the deeper stuff. Launching our canoes by paddling a "white-water" downhill ski run was especially entertaining. On the water, conditions were excellent with mostly clear skies and moderate air temperatures during the day—T-shirt weather at times for the younger people. The horizons of sparkling white mountains against solid blue skies were pure majesty. While mountain names on the map fuelled imagination, kilometres of shelf ice towering above our heads gave us constant reminder of the river's heft and dynamism.

**Endeavor Ambition Valhalla Commander
Laura Pereleshin Saddlehorn Cinema Cirque
Devil's Elbow Eagle Crag Cone Alpha Choquette**

It was wonderful. The lads enjoyed great camaraderie while learning new skills in an awesome environment. Yes, there were some bruised ribs, wet feet, and cut fingers, but there were no broken bones and no upset canoes. As an assistant to the overall exercise, I was encouraged to provide an account of resource-management issues affecting this river while sharing my concern for rivers everywhere … along with a bit of river poetry spouted around the campfire (Appendix C). Auditing student information sessions, from map reading to geology, I continued my own education while celebrating the options now available to these young people, two generations removed. Intelligent and courageous, they all seemed aware of current issues around health of the planet and the need for increased respect toward natural systems as well a renewed look at how we manage them (or not). Great glaciers and hot springs were bonus attractions for these young tourists. Watching this generation drift through ancient forests and onto the tidal flats, I found myself reflecting on the multitudes of people who must have passed this way before us—fishers and trappers, hunters and traders, miners and loggers, explorers and adventurers—their spirits alive in the surrounding trees. Fifteen years later, in composing my thoughts for this book, I found myself reflecting on the multitude of people touched by the river in more recent times: Rosemary and Irving Fox who had inspired my generation to take action; May Murray who was the heart and soul of our organization; Grant Copeland who was our irrepressible point man in the toughest of times. When honouring the anonymous donor who had started the ball rolling, I also wanted to honour all those who had taken part in the journey, especially those friends who have since passed. I didn't know there would be quite as many.

Gary Fiegehen was taken by cancer on August 26, 2022, and I have lost a dear friend. Beginning with our first canoe trip in 1982, we were good buddies and comrades-in-arms around all things Stikine: without any planning on the subject, we always seemed to be moving in the same direction, as with this book project. Insightful and honest, Gary was a pleasure to be with. His contribution to this story far exceeds his gifts of photography: his heart was in it from the start and his spirit imbues every page. The human being named Gary Fiegehen is with me always.

Sauntering … strolling … lingering … loitering
Dabbling and dawdling through a coastal estuary
Weaving and wandering through tidal-flat islands
Shifting and sifting through saltwater sandbars
Settling and surrendering, mixing and merging
The Great River and the Great Ocean are one

17
Mixing and Merging

This 2001 photograph from the home of May Murray aptly captures the mood of the day while putting a picture on the "mixing and merging" process. The Friends of the Stikine old guard was giving way to the new guard while in the company of facilitators who had made the transition possible; meantime, in the photographic moment, everyone was celebrating the recent designation of a provincial park on the upper Stikine River. This last meeting of its kind brought a sense of closure to many of us. In the back row, left to right, Ann Jacob, Gil Arnold, Stan Tomandl, Dona Reel, Ric Careless, May Murray, Maggie Paquet, and Jennifer Voss; up front, Gary Fiegehen and Peter Rowlands.

The Stikine had captured and enamoured me, and in surrendering to that attraction, I learned much about myself and about the world I live in. One of the most significant things I discovered—among a host of shortcomings—is the presence of my extreme good fortune. While lucky to survive several novice paddling events, I am fortunate to have experienced the Stikine in a near-natural state before recreational signage and industrial improvements. From there, I came to know how fortunate I am to live in a society that welcomes participation in administrative decision-making, however skewed the process might seem. I also came to truly appreciate the freedom to express my opinions and to write my story as one of many voices in a human choir. I am fortunate in having family and friends who have guided me. I am lucky to be here. I am lucky to be alive.

Although not necessarily a direct cause, my twenty-year affair with the Stikine River coincided with failures in my human relations department. Together with the loss of friendships by way of my unconsciousness during the process came the demise of marriages and romantic relationships—intensity prevailing over longevity—but that's another story. Meanwhile, my retreat from the Stikine came with a menu of other wonderful rivers to paddle in Canada, from coast to coast to coast. The stars also aligned for reconnection with my high-school sweetheart, Susan. After forty years without contact, we are again comfortable in completing each other's sentences while somehow sharing the same collection of attitudes and beliefs. I'm very lucky.

Whatever the underlying factors might have been at the time, I became hooked by the Stikine at first contact. The intentional disablement of such a magnificent stretch of running water—an archetype if ever there was one—was a scenario impossible to live with. History offers mixed reviews. Although much of the watershed and many of its features are now in parks and protected areas, it remains to be seen if the newly created special management zones will be effective in protecting the vital headwaters and in defending the lower river system against the detriments of aggressive and excessive mining initiatives. Even with an enthusiastic fish-enhancement program in effect, salmon counts on the Stikine are generally on the wane and the fishery remains subject to restrictions and closures, especially for chinook and sockeye. It seems, the once robust coho stocks of the Iskut have all but disappeared. Nevertheless, the Stikine ecosystem remains in better shape than many others and it might represent a step forward on the road toward greater awareness on a global scale.

When we Stikineers hopped on the freshwater bandwagon some forty years ago, we often feared overreacting to an issue already identified and being addressed by saner heads. Not so. Thanks to the diligence of (Saint) Maude Barlow and her Council of Canadians, we now know that a handful of transnational corporations already control a large percentage of Earth's freshwater reserves, making billions and billions of dollars by extracting it to the detriment of agriculture and selling it in plastics which are severely infecting the ocean, the fish, and even ourselves. Aquifers everywhere are drawing down and soil is drying out wherever buckets of rain are not flooding us out; air temperature is going steadily upward, and the ocean is becoming more acidic every day. How long have we got?

Although the relationship between available fresh water and climate change has yet to be quantified, the need for conserving natural ecosystems in the name of *global sustainability* has been well documented since the 1980s; since then, a shift in economic strategy has brought much higher emphasis onto profit margins and stock returns in the interests of *corporate sustainability*. Results are not encouraging. Many of the little guys are being eaten by a few big guys and everyone's perceived need to fight for survival exerts relentless pressure downward onto the planet. Environmental priority seems impossible unless industry finds a way of making money out of it.

Meanwhile, society is exploding outward at such high velocity we are either sucked along by it or left in a back eddy. Space tourism and the electrification of everything has taken precedence over economic disparities and global hunger. Recent shameful decline of the once-respected Boeing Corporation suggests the moneychangers are not only in the temple, but they now think they own it. Recent assaults on Washington and Ottawa by restless citizenry hint to the introduction of anarchy fuelled by the religious right. With a global pandemic thrown in, it is increasingly difficult to define "normal," and with the Russian invasion of Ukraine, it's easy to identify humankind as the greatest threat to planetary health.

We are a high-speed species on the move, riding sharply honed ideologies in pursuit of happiness and security based on power and wealth—rushing blindly past our goal while destroying the field that supports the game. Assuming humans are capable of coexistence and wish to inhabit a healthy planet, we have a steep ecological hill to climb, and no amount of techno-electric wizardry will take us over the top. If we want it, we must do it ourselves. We must care for our precious natural resource while forgiving those who would squander it—they know not what they do.

> ***Only when the last tree is cut down, the last fish eaten, and the last stream poisoned, will we realize that we cannot eat money.*** —Indigenous proverb

In Canada at least, kudos go out to the latest crop of urban planners and developers who are showing increased respect for running water by not paving over every last headwater stream, leaving more of them exposed to light for interaction with wildlife and humans while ensuring construction projects remain increasingly subject to watershed realities. Similarly, in Canada and elsewhere, recent initiatives to give Indigenous and other local people increased responsibility for stewardship of homeland resources provides hope. Unable to give back everything we have already taken; we might now benefit by finally agreeing to share the land. In so doing, we have no choice but to trust such initiatives will come with corresponding fiscal and administrative support rather than simply being additional downloads of expenses without meaningful control (while funds are being preserved at the top for investment in "greater" things). Given the overall picture, any glimmer of hope in any small corner of the world is worth celebrating. Garbage bags on gateposts come to mind, placed there by local citizens to collect discarded refuse from urban trail users; empty coffee cups are apparently heavier to carry than full coffee cups.

Perhaps, as once prophesized, we meek old geezers will inherit the Earth, or what's left of it, while the progressive "true believers" move along to Mars and live happily ever after, manufacturing their own water substitute. No moral judgment intended; however, to my mind, our currently determined exploration of outer space seems more about acquiring physical resources than it is about gaining knowledge. We already know enough to know our universe is a large and complex system, one we will never appreciate the full extent of or ever fully understand. There's magic in it. Of course, if I were an astrophysicist or a space-shuttle pilot, my perspective would probably be different. Nevertheless, to a tree-hugging dinosaur refusing to let go, it seems our best option is to get down to the business of cleaning up our base camp—we may have more luck attracting universal intelligence than finding it at the end of our line of space garbage. To explore is human; however, if we don't first see to the health of our home planet, we don't deserve to go beyond it. Curiosity? Greed? Whatever our motivators, humankind seems destined to reproduce and to expand outwardly as quickly as possible. Remind you of anything?

Unless I am but a fearmonger crying wolf for reasons entirely of my own imagination, the time has come for all of us to be singing the same song with the same message and to be singing it loud and clear without religious contamination, political-isms, or racial profiling.[22] Be it heaven or hell, we are in this together, and we can't expect leadership from the top. Power-tripping autocrats are everywhere, and by default or by design, they have little interest in anything outside themselves. Governments, democratic and otherwise, have largely been co-opted by the corporate state which, with few exceptions, has little interest in anything not beneficial to its bottom line. Those volunteer organizations and noble souls dedicated to the well-being of their fellow humans are barely discernable in the busy dust of economic progress, and they are at risk of falling through ever-widening cracks in our social structure.

No, we can't shut down the economic engine and write a new paradigm; however, we can direct our votes and our investments toward those most committed to respect for our natural world. Planet Earth would comfortably survive without us, but, if we want to be here, we have to honour it. Though human life brings pleasure and fulfillment, no one said it was supposed to be easy. Being human is to be involved—revisionary, not revolutionary. Rights without responsibility is tyranny. If we continue shortcutting through ecology, we will become lost. If we sit back and buy every labour-saving device on the market, we are dead.

There's no easy answer and there's no simple cure. Relentless pursuit of small-scale improvements around the globe seems the only way to go. Otherwise, sooner or later, the human future may require a choice between spaceships and fresh drinking water. Meantime, some fortunate folks such as us still have a vote to give while maintaining faith in the human journey aboard this planet. Never, never give up.[23]

Mixing and Merging

Author Notes

1. Artist's rendering of **Site Zed**, the uppermost of two dam sites planned for the Grand Canyon of the Stikine.
2. Born of mixed European—Indigenous parentage to become children of our first transnational industry, the **Métis** people of the fur trade might well be regarded as the first true Canadians.
3. **Samuel Black**—*A Journal of a Voyage from the Rocky Mountain Portage in Peace River to the Sources of Finlays Branch and North West Ward in Summer 1824*, ed. E. E. Rich and A. M. Johnson (*HBRS*, 18, London, 1955).
4. *Dictionary of Canadian Biography*—**Samuel Black** by George Woodcock.
5. A rain shadow in lee of the Coast Mountains provides a beneficial **microclimate** for the area between Telegraph Creek and the Chutine River valley, which is characterized by terraces containing classes 4 and 5 agricultural lands.
6. **Fish counts** for the annals of 1979 (1980): springs/kings = 775 (1,488); coho = 10,700 (6,500); sockeye = 10,500 (18,100).
7. My references to **Christian** lore are almost entirely from Sunday School at the United Church of Canada. Except for marriages and funerals, I have not attended many church services since childhood. Jesus, yes ... religion, no.
8. During his tenure with Southeast Alaska Conservation Council (SEACC), **Jim Bourquin** also served as a boatman for Alaska Wilderness Expeditions; he and several friends are renowned for their winter ski trip on the ice from Telegraph Creek to the river's mouth—the first such known feat in modern times. Jim remains a trusted guide for the trails and rivers of Stikine.
9. A friend from beginning to end, **Grant Copeland** began his Stikine focus with creation of a comprehensive visitor's guide (1985) before devoting much of his later life to the cause. While supporting ecological issues across a broad spectrum, he found time to transform an abandoned Selkirk Mountain mine site into an eco-friendly retreat-style lodge for mountaineers and cat skiers. Immediately prior to being taken by cancer at age sixty, Grant published a book outlining concerns and recommendations that remain relevant today. See *Acts of Balance: Profits, People and Place* (1999).

10. A pleasant aside from paddling and politicking occurred during a visit to Telegraph Creek soon after our B737 field trip when photographer Gary Fiegehen introduced me to a gentleman named **Henry Quock** who was particularly happy to meet the airplane driver and to receive confirmation of our flightseeing event. Being the only resident in the village to report seeing a large, orange eagle fly past in the very early morning of June 4, 1986, Henry was having to endure questions about the authenticity of his visions.
11. **Grant McConachie** (1909–1965): Bush pilot/owner of Yukon Southern Air Transport who became president of Canadian Pacific Airlines in 1947. See *Bush Pilot with a Briefcase: The Incredible Story of Aviation Pioneer Grant McConachie* (1972) by Ronald A. Keith.
12. **Simon Gunanoot** (1874–1933): A prosperous Gitxsan merchant of Hazelton who was accused of murder and avoided capture by living in the wild with his family for thirteen years. Eventually surrendering, he stood trial and was acquitted—his legend remains. See *Trapline Outlaw: Simon Peter Gunanoot* (2005) by David Williams.
13. **Robert Campbell**—his papers were mostly destroyed by fire in 1882. *The Discovery and Exploration of the Pelly (Yukon) River* was published in 1883 (Toronto); *Two Journals of Robert Campbell, Chief Factor, Hudson's Bay Company, 1808 to 1853* was published in 1958 (Seattle).
14. *Dictionary of Canadian Biography*: "Robert Campbell," by Kenneth Stephen Coates.
15. Excerpts from Campbell's papers by R. M. Patterson in his book, *Trail to the Interior* (1966).
16. **Crowsnest Pass and Tumbler Ridge** are separate coal mining operations in the Rocky Mountains that received provincial government support while producing considerable environmental concern for little return to the public purse.
17. Although my panic-mode need for an informational film about Stikine abated with BC Hydro's withdrawal from the canyon, the movie-making idea never really went away. At close of the last century, I went into partnership on a camera-mounted T-33 (T-Bird) jet trainer in hopes of flying my own movie. Alas, the airplane was too expensive to maintain, and Raven Air Photo disappeared as quickly as it had appeared. Unable to kick the habit, my fantasy was later realized in **Stikfliks**—airborne Stikine River home movies constructed from animated still photographs and set to music. The end.
18. **Forest grazing permits** are used by livestock producers to increase the quantity of pasture, temporarily or permanently, while helping protect livestock against the elements.
19. Relocating from Montana in the 1970s, **Mark Angelo** was known as "Mr. River" while spearheading numerous river conservation efforts in the greater Vancouver area. As longtime head of the Fish, Wildlife and Recreation Program at BCIT and rivers chair at ORC, Mark founded and chaired BC Rivers Day and World Rivers Day as well as taking leadership roles with both the British Columbia Heritage Rivers System and Canadian Heritage Rivers System. Having participated globally in several river-conservation films, Mark has received honorary doctorates from Simon Fraser University and from Trent University, as well as being inducted into the Order of British Columbia and into the Order of Canada.
20. *The Stikine River* (1979) published by the **Alaska Geographic Society** contains a comprehensive list of boats used on the river along with their descriptions and photographs.

21. **Eric Buss,** aged 40, died the following year (November 27, 1991), taken by an avalanche on Hudson Bay Mountain, near Smithers, while training with the local Volunteer Mountain Rescue Squad.
22. Only **one** human race is known to inhabit this Earth.
23. "Forgive your parents and believe in yourself. Pursue your God-given talents and never, never give up"—encouraging words from **Peter D. Friesen** who escaped Ukraine and the tyranny of post-revolutionary Russia to receive an elementary school education on the Canadian prairie before inventing the machine that helped him become a world leader in the risky business of relocating hotels, lighthouses, and other large man-made structures. See *Man on the Move: The Pete Friesen Story* (2009).
24. Because of Gary's sudden passing, I am unable to properly confirm the location for several photographs which have been identified as such in his **photography notes.**
25. A concluding **photo montage** features people of the Stikine who contributed in some way to my experience, whether or not referenced in the narrative. The inset aircraft image *(#720 Empress of Stikine)* is copied from a full-colour representation created by artist Dan Fallwell who was a CP Air customer service agent in BC District at the time.

Appendix A
Voices Passed

- Irving Fox
- Rosemary Fox
- Bill Mason
- Grant Copeland
- Colleen McCrory
- May Murray
- Murray Wood
- Ron Janzen
- Johnny Tashoots
- Dan Pakula
- Mike Jones
- Bill Murray
- Eric Buss
- Francis Gleason
- Tannis Fisher
- Gordon Hartman
- Garry Grant
- Edward Rowlands
- Tom Buri
- Gary Fiegehen

Appendix B
Declaration of the Tahltan Tribe, 1910

We, the undersigned members of the Tahltan tribe, speaking for ourselves and our entire tribe, hereby make known to all whom it may concern that we have heard of the Indian Rights movement among the Indian tribes of the Coast and of the southern interior of B.C. Also, we have read the declaration made by the chiefs of the southern interior tribes at Spences Bridge on the 16th of July last, and we hereby declare our intention to join with them in the fight for our mutual rights, and that we will assist in the furtherance of this object in every way we can until such time as all these matters of moment to us are finally settled. We further declare as follows:

Firstly—We claim the sovereign right of all the country of our tribe … this country of ours which we have held intact from the encroachments of other tribes, from time immemorial, at the cost of our own blood. We have done this because our lives depended on our country. To lose it meant we would lose our means of living, and therefore our lives. We are still, as heretofore, dependant for our living on our country, and we do not intend to give away title to any part of same without adequate compensation. We deny the B.C. government has any title or right of ownership in our country. We have never treated with them, nor given them any such title.

Secondly—We desire that a part of our country consisting of one or more large areas (to be selected by us) be retained by us for our own use … said lands and all thereon to be acknowledged by the government as our absolute property. The rest of our tribal land we are willing to relinquish to the B.C. government for adequate compensation.

Thirdly—We wish it known that a small portion of our lands, at the mouth of the Tahltan River, was set apart a few years ago by Mr. Vowell as an Indian reservation. These few acres are the only reservation made for our tribe. We may state we never applied for the reservation of this piece of land, and we had no knowledge why the government set it apart for us, nor do we know exactly yet.

Fourthly—We desire that all questions regarding our lands, hunting, fishing, etc., and every matter concerning our welfare be settled by treaty between us and the Dominion and B.C. governments.

Fifthly—We are of the opinion it will be better for ourselves, also better for the governments and all concerned if these treaties are made with us at a very early date, so all friction and misunderstanding between us and the whites may be avoided, for we hear lately much talk of white settlement in this region and the building of railways, etc., in the near future.

Signed at Telegraph Creek, B.C., this eighteenth day of October, nineteen hundred and ten by Nanok (Chief of the Tahltans), Nastulta (Little Jackson), George Assadza Kenetl (Big Jackson), along with eighty other members of the tribe.

Appendix C

RIVERS OF CANADA

PEACE ATHABASCA ABITIBI MIRAMICHI
RESTIGOUCHE RICHELIEU ASSINIBOINE NECHAKO
KLUANE NAHANNI NIPIGON MISSINAIBI
NANOOK NOTTAWAY NOTTAWASAGA NASS

melting of ancient ice
the rivers of Canada
rise on sacred ground

OKANAGAN NIAGARA OMINECA NASKAUPI
KOOTENAY COWICHAN NATASHQUAN SQUAMISH
KITIMAT KAPUSKASING NITINAT TEMISCAMIE
CHILCOTIN NANAIMO TAHLTAN SIKANNI

dancing to many drums
the rivers of Canada
celebrate one and all

MISTASSINI TEMAGAMI KIPAWA KATTAWAGAMI
TATSHENSHINI CHICOUTIMI RICHIBUCTO RIMOUSKI
SKEENA KECHIKA KITLOPE KHUTZEYMATEEN
BELLA COOLA BABINE SPATSIZI STIKINE

shining in many eyes
the rivers of Canada
welcome one and all

JACQUES CARTIER SAGUENAY CHAMPLAIN ST. LAURENT
OTTAWA MATTAWA WINNIPEG SASKATCHEWAN
YUKON COLUMBIA CHURCHILL MACKENZIE
FRASER THOMPSON FINLAY FRANCES DEASE

singing in many tongues
the rivers of Canada
speak to one and all

SHELBURNE SHUBENACADIE KANANASKIS QUESNEL
RED DEER YELLOWKNIFE COPPERMINE QU'APPELLE
MUSKOKA MADAWASKA MISSISSAUGA SAUGEEN
VICTORIA VANCOUVER MONTREAL ST. JOHN

ancient providers and highways of commerce
the rivers of Canada
are the bloodlines of our land

Gary Fiegehen
Photography Notes

Front Cover — **Stikine River,** upstream view of the lower river near Great Glacier where oral tradition tells of a time when the river coursed through a tunnel of ice caused by the merger of what are now Great Glacier and Choquette Glacier through which four elders volunteered for the first attempt which proved successful; the fact that several diverse populations of Indigenous peoples along the coast all speak of a migration from the great river suggests many others once passed through that tunnel of ice.

Dedication Page — **Spectrum Range** north to the summit of Mount Edziza atop a volcanic complex measuring 24 kilometres wide, 64 kilometres long, and 2 kilometres thick which, according to plate tectonics theory, was formed 10 to 15 million years ago when America drifted westward to override its neighbour with extreme fire and brimstone, including numerous floods of lava. Concurrent with the Cordilleran ice sheet (which retreated about 10,000 years ago), the main vent of Edziza became sealed by its own lava plug while volcanic eruptions continued from surrounding vents, cones, and fissures.

VIII Preface — **Mount Edziza,** caldera-edge comb at 2780 metres (9121 feet) above sea level—the geophysical centre of the Stikine watershed.

XI Preface — **Mount Edziza,** the eastern shoulder of the volcanic complex with a portion of its resident glacier.

Page 10 — **Retreating glacier** in the high country spawning fresh water in symmetric fashion.

Page 23 — **Eaglenest Range** at the southwest corner of Spatsizi Plateau, thought to create a rain shadow favourable for caribou habitat.

Page 24 — **Stikine headwaters,** our "GP Creek" at the foot of "Mount Marsden" on its way to Tuaton Lake.

Page 33 — **Osborn Caribou,** commonly seen on the high-country plateaus of Spatsizi and Edziza where winter winds remove snow cover from their necessary forage of lichen and grass.

Page 34 — **Cold Fish Lake** emptying into Mink Creek flowing east toward Spatsizi River. At mid-frame left is Gladys Creek flowing in from Gladys Lake Ecological Reserve.

Page 35 — **Spatsizi Plateau** and one of many unnamed tributaries contributing to Stikine.

Page 47 — **Sergief Island** in Dry Strait pointing east toward mouth of the Stikine River—where invertebrate life abounds, where grasses and willow bind flour-fine silt into sod and create an environment which sustains a wide variety of indigenous and migratory birds in very large numbers.

Page 48 — **Kitsu Peak,** "northern lights" (aurora borealis) in the Tahltan language, a typically colourful remnant of a shield volcano in the southern Spectrum Range.

Page 57 — **Telegraph Creek** with the RiverSong Café and General Store prominent on the riverfront and the narrow cut of the namesake creek containing the only road up to the main village.

Page 58 — **Site Zed,** the uppermost dam site in the Grand Canyon of the Stikine—the white marker at upper-left indicates the envisioned depth of impoundment water behind it. Fortunately, today, it remains about as naturally wild as a river can be.

Page 70 — **Level Mountain,** a remnant volcanic complex more than 15 million years old (twice the age of Edziza) located north of Stikine's central canyon and part of the same Stikine volcanic belt—one small section of the circum-Pacific Rim of Fire which stretches from the tip of South America around to New Zealand. Here, Level Mountain is a large part of the divide between the Stikine–Tuya and the Nahlin–Taku rivers to the west and the Teslin–Yukon rivers to the north. From this section of the Continental Divide, waters also flow directly east into the Dease–Liard–Mackenzie river systems.

Page 90 — **Pack train** at the edge. Extended trail rides are available from local guide outfitters.

Page 91 — **Canyon goats,** a special breed of about 300 bearded and dagger-horned animals that climb down from the trees for safety—North America's sole representative of the small *rupicaprid,* or goat–antelope family, related to Europe's chamois and Asia's goral. Long hair, woolly undercoats, and flexible traction-pad hooves are standard equipment.

Page 92 — **Meditating goats** (the second goat's white rump is barely discernable) contemplating the words of Wendell Berry from *The Unsettling of America* where he cautions against the severance of man from nature: "The threat is not only in the totalitarian desire for absolute control. It lies in the willingness to ignore an essential paradox: the natural forces that so threaten us are the same forces that preserve and renew us." More fragile than they appear, canyon walls are vulnerable to water intrusion which separates slabs and sends them downward at great speed and distance.

Page 93 — **Tanzilla Gap** with a human observer on one side and high-water driftwood on the other.

Page 96 — **Tahltan Flat** at the confluence of the Tahltan (foreground) and Stikine (entering frame-left)—traditional fish-harvesting centre for the Tahltan people and site of myriad summer rendezvous with their Tlingit cousins and other trading partners. The road to Telegraph Creek climbs past cliffs of layered basalt which, according to local lore, represent Raven's groundhog blanket with its two distinct colours referencing the two kinds of groundhog present in the territory.

Page 105 — **Unknown,** perhaps Spatsizi River near Hyland Post.

Page 106 — **Happy moose** in their natural habitat.

Page 117 — **Great Glacier** coming off the Stikine Icefield at Kate's Needle, slightly right of centre, and flowing some fifteen kilometres past Mt. Pratt before feeding its lake and river which empties into the Stikine beneath the cloud at frame-left where a BC provincial park now provides camping and a viewing trail for the glacier. Boots and woollies recommended.

Page 118 — **Great Glacier** backdrop to lower river fishers.

Page 125 — **Mount Klappan** near the southeastern divide with Little Klappan River flowing northwest in a broad valley known as prime caribou calving grounds and for its coal mining potential. The headwaters of the Spatsizi and Stikine rivers are slightly out of frame to the left, and headwaters of the Nass and Skeena rivers are in the mountains at top right—Sacred Headwaters.

Page 126 — **Hudson's Bay Flat,** in the foreground on river-right at Glenora where perhaps as many as 10,000 stampeders wintered enroute to the Klondike gold rush in 1897–98 before many changed their minds and went home. Some didn't make it either way—site of the HBC store before it was moved upstream to Telegraph Creek where, many decades later, it morphed into the RiverSong Café and General Store. The mountains in the background of the photograph are locally known as the "Stingies" for their low harvest rate of edible ungulates while forming the northeast corner of the high elevation "Golden Triangle" between the Iskut and Stikine rivers.

Page 138 — **Twin glaciers** from the 1200-square-kilometre Stikine Icefield which drapes over the Coast Range divide and the international boundary while punctured by 3000-metre-high granite peaks.

Page 139 — **Unknown,** most likely a mining exploration camp somewhere in the golden triangle.

Page 140 — **Unknown,** probably a Dall or Stone sheep in the Edziza complex.

Page 141 — **Unknown** mining camp, perhaps Skyline's operation on Johnny Mountain.

Page 153 — **Lower Stikine,** supporting five species of salmon along with moose, eagles, and untold other species while winding through forests of cottonwood and spruce—a dynamic eco-region in delicate balance and at the mercy of modern industrialism.

Page 154 — **Smokehouse** of Tahltan village where a storied girl returns after living with the salmon people to tell her parents: "You must treat the salmon people respectfully; you must never talk evil of them, nor disparage them or their flesh. If you do not heed these things, they will take revenge on you."

Page 159 — **Unknown,** meltwater tarn in source waters of the Iskut River system.

Page 160 — **Glacial meltwater** tumbling down Hoodoo Mountain enroute to the nearby Iskut River.

Page 169 — **Lower watershed,** a mother moose and her calf stroll through the neighbourhood.

Page 170 — **Great Glacier,** calving bergs into its lake melting into its river feeding Stikine, reminding us of the world's freshwater reserves largely being stored in icecaps remaining from the latest ice age—not an infinite resource. Each polluted stream moves us further into deficit.

Page 179 — **Stikine River estuary** at low tide opening south-southwest toward the "goose-head" northern tip of Wrangell Island with Rothsay Point at mid-upper left and Sergief Island near right.

Page 180 — **Stikine Strait** southwest toward Zarembo Island. Wrangell is slightly out of frame left; Petersburg is farther out of frame top-right.

Page 185 — **Mount Edziza** at top of the watershed displaying a section of its complex geology.

Page 186 — **Spectrum Range** with extreme colouration resulting from expelled gases trapped beneath the ice during eruptive events and here exhibiting a degree of "plasticity."

Page 190 — **Moonrise** over Mount Edziza.

Page 192 — **Willie Williams and Stormy** glassing for goats.

Page 194 — **Unknown,** perhaps the middle Iskut River at Snippaker Creek.

Page 196 — **Eve's Cone,** on the northwest shoulder of the Edziza complex, named for Eve Brown who survived an avalanche that claimed her companions while travelling from Telegraph Creek to Iskut by dogsled in 1950.

Page 201 — **Iskut Headwaters,** one of numerous on the eastern flank of the Edziza complex.

Page 204 — **Tlingit totem** in downtown Wrangell.

Back Cover — **Happy Valley** above Tuaton Lake being admired by GP in 1982.

Acknowledgements

This is a book about the work of many people. As well as thanking every person named in the story, I wish to acknowledge the legions of unnamed souls who supported them, directly or indirectly, during two decades of journeys and meetings. Also vital were colleagues and committee members on both sides of the issues table—earnest people and diverse ideas create interesting conversations. In retrospect, personal solace comes from knowing my worst mistakes often made the best stories. Thank you, dear people.

For this story, a special nod goes to the late Bill Mason for sharing paddling techniques that have made the telling possible, to Wade Davis for inspiration in completing journeys within and without, and to the late Dan Pakula for decades of guidance on the river and beyond.

I give great thanks to Ann Jacob and Stan Tomandl at Friends of the Stikine Society for their donation of press clippings that prompted long-held musings into dedicated effort; while thanking the anonymous donor of those press clippings, I offer hearty thanks to Tom Buri for igniting the process with workshop reports from the distant past. I am especially grateful to my longtime friends and colleagues—editor Maggie Paquet and photographer Gary Fiegehen—who helped shape this vessel and send it on its way. Their spirits prevail.

For help with details of the text, I thank Diane Pakula, Tannis Fisher, David Fisher, Barb McCutcheon, Rick McCutcheon, and Cheryl Reitz in Telegraph Creek, as well as Ray and Reg Collingwood in Spatsizi; for river-specific issues, I thank Bill Sampson, Travis Pakula, Tony Shaw, and Neil Piller; for updated protected area status and nomenclature, I thank Nicole Hepp and Andrew Patrick at BC Parks; for information about the current health of non-human inhabitants of the watershed I thank Chris Zimmer at Rivers Without Borders.

For producing area maps (which I consider essential), I give thanks to Hap Wilson for creative direction and to Abby Reynolds for creative polish. Their finished products are pleasingly simple and appropriate to the text.

For her editorial expertise in both words and direction I warmly thank Janet Gyenes; for their design and support in making this story a published reality I thank the people of FriesenPress. For assistance in cleaning up loose ends and bringing closure, I give big thanks to my friend and fellow scribe, Gary Johnson.

For everything above and beyond, I give deepest thanks to my partner in life, Susan Bright, who has been with me on every step of this publishing path—I am honoured and blessed.

This book is a personal undertaking, and I am responsible for all errors, omissions, and opinions.

PER.21Dec22

Gary Fiegehen
(1947–2022)

A Ryerson trained photographer introduced to the Stikine River in 1982, Gary Fiegehen had since continued to focus on the people and landscapes of British Columbia, publishing seven books and contributing to numerous others, most notably his *STIKINE: The Great River* (1991). He has contributed to environmental campaigns and worked with various First Nations of B.C., including the Nisga'a during their successful efforts to attain the first modern-day treaty in the province. Having seen a great deal of his adopted province through his lens, Gary considered the Stikine watershed to be a very special landscape and he was grateful to the author for their introductory canoe trip. Until his passing, Gary lived in Vancouver with Sara, his best friend and motivator. This selfie is from a solo trip on the Spatsizi–Stikine in 1986.

Peter Rowlands
(1944 –)

A career aviator, Peter Rowlands was an avid paddler and skier until artificial leg joints helped transform him into a dedicated cyclist and storyteller. He now lives north of Toronto with his best friend and reluctant critic, Susan. This photograph from Metsantan Canyon was taken by his father in 1989.

Ω

Printed in the USA
CPSIA information can be obtained
at www.ICGtesting.com
JSHW061052140823
46476JS00002B/16